Lecture Notes in Mathematics

Edited by A. Dold and B. Eckmann
Series: Australian National University, Canberra
Advisors: L. G. Kovács, B. H. Neumann, M. F. Newman

680

Mathematical Control Theory

Proceedings, Canberra, Australia,
August 23 – September 2, 1977

Edited by W. A. Coppel

Springer-Verlag
Berlin Heidelberg New York 1978

Editor
W. A. Coppel
Department of Mathematics
Institute of Advanced Studies
Australian National University
Canberra, ACT 2600/Australia

Library of Congress Cataloging in Publication Data

Conference on Mathematical Control Theory, Australian
   National University, 1977.
   Mathematical control theory.

   (Lecture notes in mathematics ; 680)
   Includes bibliographies and index.
   1. Control theory--Congresses. I. Coppel, W. A.
II. Title. III. Series: Lecture notes in mathematics
(Berlin) ; 680.
QA3.L28   no. 680   [QA402.3]   510'.8s [629.8'312]
                                              78-11960

AMS Subject Classifications (1970): 93XX, 49XX

ISBN 3-540-08941-1 Springer-Verlag Berlin Heidelberg New York
ISBN 0-387-08941-1 Springer-Verlag New York Heidelberg Berlin

This work is subject to copyright. All rights are reserved, whether the whole or part of the material is concerned, specifically those of translation, reprinting, re-use of illustrations, broadcasting, reproduction by photocopying machine or similar means, and storage in data banks. Under § 54 of the German Copyright Law where copies are made for other than private use, a fee is payable to the publisher, the amount of the fee to be determined by agreement with the publisher.

© by Springer-Verlag Berlin Heidelberg 1978
Printed in Germany

Printing and binding: Beltz Offsetdruck, Hemsbach/Bergstr.
2141/3140-543210

PREFACE

These Proceedings provide a record of the invited lectures delivered at a Conference on Mathematical Control Theory which was held at the Australian National University, Canberra from August 23 to September 2, 1977.

Control theory lends itself to mathematical analysis more than many other branches of science. In convening the conference my intention was to bring together Australian mathematicians working in this area and to attract others to it.

The conference was attended by more than 70 people. Courses of 5-8 lectures were given by Professors K.J. Åström, R.W. Brockett, H. Halkin, R.E. O'Malley, and R.T. Rockafellar. One-hour survey lectures were given by Dr P.E. Kloeden, Professor I. Kluvanek, and Professor J.B. Moore. There was also a programme of contributed half-hour talks, only the titles of which are reproduced here.

I am grateful to the Research School of Physical Sciences, Australian National University for financial support; to the Conference Secretary, Dr A. Howe for invaluable help; to Mrs Barbara Geary, who typed all the contributions except that of Professor Åström; and to Dr M.F. Newman for editorial assistance.

W.A. Coppel

TITLES OF CONTRIBUTED LECTURES

Russel Cooper (Department of Economics, Monash University)
    Intertemporal duality: application to consumer theory

W.A. Coppel (Department of Mathematics, Institute of Advanced Studies, Australian National University)
    Exponential dichotomies

B.D. Craven (Department of Mathematics, University of Melbourne)
    Lagrangean conditions and quasiduality

G.C. Goodwin (Department of Electrical Engineering, University of Newcastle)
    Estimation in closed loop

S.A. Gustafson (Computer Centre, Australian National University)
    A parabolic boundary-value control problem with possible applications in the steel industry

Paul Kabaila (Department of Electrical Engineering, University of Newcastle)
    A formulation of estimation and structure choice problems based on ultimate use

I.B. Lennard (Department of Mechanical Engineering, University of Western Australia)
    Iterative instrumental variables as applied to a multiple input system

B. Molinari (Department of Computer Science, Australian National University)
    Strong reachability and linear system structure

P.J. Moylan (Department of Electrical Engineering, University of Newcastle)
    Linear system inversion using differential operator notation

ADDRESS LIST OF INVITED SPEAKERS

K.J. Åström: Department of Automatic Control,
                    Lund Institute of Technology, Lund 7, Sweden.

R.W. Brockett: Division of Engineering and Applied Physics,
                    Harvard University, Cambridge, Massachusetts 02138, USA.

H. Halkin: Department of Mathematics,
                    University of California, San Diego,
                    La Jolla, California 92093, USA.

P.E. Kloeden: School of Mathematical and Physical Sciences,
                    Murdoch University, Murdoch, WA 6153, Australia.

I. Kluvanek: School of Mathematical Sciences,
                    Flinders University, Bedford Park, SA 5042, Australia.

J.B. Moore: Department of Electrical Engineering,
                    University of Newcastle, Newcastle, NSW 2308, Australia.

R.E. O'Malley: Department of Mathematics,
                    University of Arizona, Tucson, Arizona 85721, USA.

R.T. Rockafellar: Department of Mathematics,
                    University of Washington, Seattle, Washington 98195, USA.

CONTENTS

K.J. ÅSTRÖM
    Stochastic control problems   1

Roger W. BROCKETT
    Lie theory, functional expansions, and necessary conditions in singular optimal control   68

Hubert HALKIN
    Necessary conditions for optimal control problems with differentiable or nondifferentiable data   77

P.E. KLOEDEN
    General control systems   119

Igor KLUVÁNEK and Greg KNOWLES
    The bang-bang principle   138

John B. MOORE
    Statistical filtering   152

R.E. O'MALLEY, JR.
    Singular perturbations and optimal control   170

R.T. ROCKAFELLAR
    Duality in optimal control   219

STOCHASTIC CONTROL PROBLEMS

K.J. Åström

TABLE OF CONTENTS

Chapter 1:	Introduction	3
Chapter 2:	Minimum Variance Control	7
	1. Introduction	7
	2. Mathematical Models	7
	3. Optimal Prediction	10
	4. Minimum Variance Control	13
	5. Applications	23
	6. References	24
Chapter 3:	Linear Quadratic Gaussian Control	25
	1. Introduction	25
	2. Mathematical Models	25
	3. Kalman Filtering and Prediction	27
	4. Optimal Control	31
	5. Comparison with Minimum Variance Control	37
	6. Applications	38
	7. References	38
Chapter 4:	Control of Markov Chains	40
	1. Introduction	40
	2. Mathematical Models	40
	3. Optimal Filtering	41
	4. Optimal Control	42

This work was partly supported by the Swedish Board of Technical Development under Contract No. 76-3804.

	5. An Example	45
	6. References	47
Chapter 5:	Nonlinear Stochastic Control	48
	1. Introduction	48
	2. Mathematical Models	48
	3. Optimal Filtering	50
	4. Optimal Control	52
	5. Linear Systems with Random Parameters	53
	6. References	60
Chapter 6:	Self-Tuning Regulators	61
	1. Introduction	61
	2. Mathematical Model	61
	3. A Simple Self-Tuning Regulator	62
	4. Analysis	65
	5. Conclusions	67
	6. References	67

CHAPTER 1 - INTRODUCTION

The purpose of these lectures is to present some basic stochastic control problems and to present mathematical theory that is useful in solving the problems. To provide a red line in the lectures they are focused on a specific problem, namely to understand feedback mechanisms which is a fundamental problem of control engineering.

A schematic picture of a process with feedback is shown in Fig. 1. The process is characterized by *inputs*, i.e. variables which can be manipulated, *outputs*, i.e. variables that can be measured, and *disturbances*. The disturbances describe the interaction between the environment and the process. It is assumed that this interaction is such that the environment influences the process but that the process does not influence the environment. The feedback mechanism receives information about the process and the environment through the measurements and it generates appropriate control actions so that the closed loop system behaves appropriately in spite of the disturbances from the environment. A common example of a feedback law is the PI regulator which is described by

$$\begin{cases} u(t) = u_{ref}(t) + K\left[e(t) + \frac{1}{T}\int^{t} e(s)\, ds\right] \\ e(t) = y_{ref}(t) - y(t), \end{cases} \tag{1}$$

where t is time, u is the input signal, y the output signal, u_{ref} and y_{ref} are reference values for the input and the output. It is very fortunate for the control engineer that many processes can be controlled very successfully using a PI-regulator provided that the parameters K and T are chosen appropriately. This fact is of course less fortunate for the control theoretician.

Feedback processes were first explored purely empirically in connection with technical systems like centrifugal governors and electronic amplifiers. It has later been found that feedback processes also play an important role in economical, biological, environmental, and social systems.

Many attempts have been made to develop mathematical theory which will help to understand and to design feedback systems. Classical control theory was largely analytical in its nature. It gave tools for analysing a given feedback system. There was a great emphasis on stability theory. Synthesis and design problems were dealt with by repeated analysis. Over the past 30 years theory which aims directly

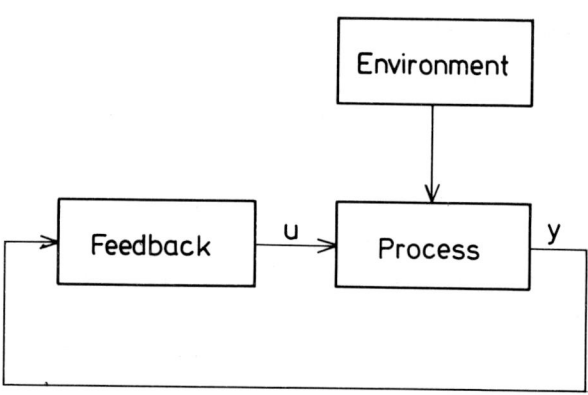

Fig. 1 - Schematic diagram of
a feedback system

at design and synthesis has been formulated. Optimal control theory is one idea. It answers the following problem. Given a description of the process to be controlled and a criterion which characterizes the desired behaviour of the closed loop system. Find the best feedback. One drawback with optimal control theory is that it does not necessarily give a feedback solution. This is shown by the following example.

EXAMPLE 1.1
Consider a process described by

$$\frac{dy}{dt} = u$$

with initial conditions

$$y(0) = a .$$

Assume that it is desirable to control the process in such a way that the performance of the system evaluated by the criterion

$$J = \int_0^\infty [y^2(t) + u^2(t)] \, dt$$

is as small as possible. It is easy to show that $J \geq 2a^2$ and that the minimum is achieved for the control signal

$$u(t) = - y(0) e^{-t}, \qquad (1.1)$$

or for the feedback law

$$u(t) = - y(t), \qquad (1.2)$$

or for any combination like

$$u(t) = -\alpha y(0) e^{-t} - (1-\alpha) y(t), \qquad \alpha \leq 1.$$

The control signal given by (1.1) is called a control program or an open loop solution because it requires only the knowledge at the measured output at time $t = 0$. Equation (1.2) gives a proper feedback law because the value of the control signal at time t is a function of the measured output at time t. It is clear that the solution (1.2) is more robust than the solution (1.1) because it will give a smaller value of the loss function if there are perturbations in the description of the model.

□

The example shows clearly that in order to get a feedback solution to an optimal control problem it is necessary to introduce disturbances and uncertainties in process descriptions. In stochastic control theory the disturbances are described as stochastic processes. Stochastic control theory will give valuable insights into the properties of feedback systems. It will give the structure of optimal feedback laws and it will e.g. tell when it is motivated to use a feedback law that is more complicated than a PI regulator. It will also in some cases give practical design tools.

From a mathematical point of view stochastic control theory is a combination of the theory of stochastic processes with the theory of differential and difference equations, calculus of variations, and optimal control theory.

The purpose of these lectures is to give an exposé of some ideas in stochastic control theory. The material for the lectures has been chosen in order to give some feel for the variety of the theory. Insight into the nature of feedback processes has been chosen as a unifying theme. For simplicity discrete time systems are treated throughout. Section 2 deals with a linear problem minimum variance control. The main virtue of this problem is that the theory is very simple and the ideas transparent. The feedback laws obtained are linear. It is thus a good starting point. The models used to describe the process and its environment are controlled ARMA (autoregressive moving average) processes or CARMA processes for short. The theory thus has strong ties to parametric time series analysis. The criterion is to minimize the variance of process outputs. Minimum variance control is of interest for control of some industrial processes where the purpose is to keep certain important quality variables as close as possible to prescribed limits. The theory will tell when and why it is useful to use a feedback law that is more complicated than the PI-

-regulator. The theory will also show the close relationships between minimum variance control and optimal predictor theory. The models used in the minimum variance control theory also occur in macroeconomics. There the models are referred to as "the reduced form" of the equations describing a macroeconomy.

The process models used in Chapter 2 are pure external descriptions. In Chapter 3 the linear stochastic control problem is approached from a different point of view. The main difference is that the model of process and its environment are now characterized by internal descriptions or state models. The criterion is again to minimize the expected value of a quadratic form. The problem statement is somewhat more general than the minimum variance problem. The major results of this theory are the Kalman filtering problem and the so called separation theorem or certainty equivalence problem which again will give important insights into the nature of the feedback control problem.

The feedback laws obtained in Chapters 2 and 3 are all linear feedbacks. In Chapter 4 we turn to models that will give nonlinear feedback laws. To keep the mathematics simple a problem with finite states is discussed. The nonlinear problem can then be solved and the solution will provide valuable insight into the properties of feedback control.

In Chapter 5 the results of Chapter 4 are generalized. The problem formulation will e.g. include the problems discussed in Chapter 2 with the additional complication of the process models now being unknown. The analysis will lead to discussion of notions of dual control, certainty equivalence, caution, and probing.

The control laws obtained in Chapter 5 are extremely complex. They can not be implemented with computing power available today. In Chapter 6 we therefore discuss simplifications that will have nice asymptotic properties. This leads to the notion of self-tuning regulators.

The books [1] and [2] listed below are useful supplementary reading. A reader interested in the continuous time problems can consult [3].

References

[1]   H Kushner: Introduction to Stochastic Control. Holt, Rinehart and Winston, New York 1971.

[2]   K J Åström: Introduction to Stochastic Control Theory. Academic Press, New York 1970.

[3]   W H Fleming and R W Rishel: Deterministic and Stochastic Optimal Control. Springer-Verlag, New York 1975.

CHAPTER 2 - MINIMUM VARIANCE CONTROL

1. INTRODUCTION

A very simple stochastic control problem is discussed in this section. It is assumed that the process dynamics can be described by an external description in the form of a difference equation with constant coefficients and that the disturbances can be characterized as ARMA processes. It is assumed that the purpose of the control is to find a feedback law such that the fluctuations in the process output are as small as possible as measured by the output variance. The mathematical models for the process, its environment, and the criterion are discussed in Section 2. It turns out that there is a close relationship between minimum variance control and optimal prediction. The prediction problem being somewhat simpler is therefore first discussed in Section 3. The minimum variance problem is then formulated and solved in Section 4.

2. MATHEMATICAL MODELS

The mathematical models used to describe the process dynamics and its environment will now be discussed. Single-input single-output systems are first treated. It is found that a generic model called a CARMA process (Controlled ARMA process) can be obtained. The multivariable version of this process is then given.

Process Dynamics

Consider a system described by Fig. 1. Assume that there is one input and one output only. It is assumed that the relation between the measured output y and the control variable u can be described by the difference equation

$$y(t) + a_1' y(t-1) + \ldots + a_m' y(t-m) = b_0' u(t-k) + \ldots + b_m' u(t-k-m)$$

This is the case for example if the process can be described by an ordinary linear differential equation with constant coefficients and a time delay and if the input signal is assumed constant over sampling intervals of unit length. Introduce the backward shift-operator q^{-1} and the polynomials

$$A_1(q^{-1}) = 1 + a_1' q^{-1} + \ldots + a_m' q^{-m}$$
$$B_1(q^{-1}) = b_0' + b_1' q^{-1} + \ldots + b_m' q^{-m}.$$

The model can then be written as

$$y(t) = \frac{B_1(q^{-1})}{A_1(q^{-1})} u(t-k). \qquad (2.1)$$

This model is often a reasonable approximation of many engineering processes that are being operated close to equilibrium conditions.

The Environment

It is assumed that the action of the environment on the process can be described by a disturbance n acting on the output. Adding a disturbance n to the output y of (2.1) gives

$$y(t) = \frac{B_1(q^{-1})}{A_1(q^{-1})} u(t-k) + n(t). \qquad (2.2)$$

There may in fact be many different disturbances acting on the process. Under the linearity assumption it is possible to use the superposition principle to reduce all disturbances to an equivalent disturbance n on the output. The disturbance n thus has physical interpretation as the output that would be observed if there is no control i.e. u = 0. Moreover it is assumed that the disturbance n can be represented by

$$n(t) = \frac{C_1(q^{-1})}{A_2(q^{-1})} \varepsilon(t), \qquad (2.3)$$

where $\{\varepsilon(t), t = 0, \pm 1, \pm 2, \ldots\}$ is a sequence of independent normal random variables and $C_1(q^{-1})$ and $A_2(q^{-1})$ are polynomials in the backward shift operator. Such a representation is certainly possible if n is a stationary stochastic process with a rational spectral density. The representation (2.3) will however not necessarily require a stationarity assumption. Non-stationary processes can be handled by letting the polynomial $A_2(\xi)$ be unstable i.e. have zeros inside the unit disc.

The CARMA Model

A combination of the equations (2.2) and (2.3) gives the following description of the process

$$y(t) = \frac{B_1(q^{-1})}{A_1(q^{-1})} u(t-k) + \frac{C_1(q^{-1})}{A_2(q^{-1})} \varepsilon(t).$$

By introducing the polynomials $A = A_1 A_2$, $B = B_1 A_2$, and $C = C_1 A_1$, this description can be simplified to

$$A(q^{-1}) y(t) = B(q^{-1}) u(t-k) + C(q^{-1}) \varepsilon(t), \qquad (2.4)$$

where

$$A(q^{-1}) = 1 + a_1 q^{-1} + \ldots + a_n q^{-n}$$
$$B(q^{-1}) = b_0 + b_1 q^{-1} + \ldots + b_n q^{-n}, \qquad b_0 \neq 0$$
$$C(q^{-1}) = 1 + c_1 q^{-1} + \ldots + c_n q^{-n}.$$

There is no loss in generality in assuming that all polynomials are of degree n because we can always put trailing coefficients equal to zero.

The mathematical model (2.4) will be called a CARMA (controlled ARMA) process, because without the control i.e. u = 0 the model is identical to the ARMA process which is commonly used in time series analysis. Notice also that without the disturbance the model is a simple rational transfer function model which is commonly used in engineering.

Notice that it is always possible to assume that the polynomial $C(\xi)$ has all its zeros outside the unit disc or on the unit circle. This is seen as follows. The polynomial $C(q^{-1})$ only enters the system description in the description of a disturbance

$$v(t) = C(q^{-1}) \varepsilon(t)$$

The signal v is completely characterized by its covariances.

$$r_v(k) = E\, v(t) v(t+k) = \sum_{i=0}^{n-k} c_i c_{i+k}.$$

The covariance $r_v(k)$ is also given as the coefficient of the term q^k or q^{-k} in the Laurent series of the function $C(\xi) C(\xi^{-1})$. But by factoring the polynomial and sorting the factors differently it is always possible to find a polynomial $\tilde{C}(\xi)$ with all zeros outside the unit disc or on the unit circle such that

$$C(\xi) C(\xi^{-1}) = \tilde{C}(\xi) \tilde{C}(\xi^{-1}).$$

A simple example will serve as an illustration.

EXAMPLE 2.1
Consider the random process $\{v(t)\}$

$$v(t) = \varepsilon(t) + c\varepsilon(t-1) = (1 + cq^{-1}) \varepsilon(t),$$

where $c > 1$ and $\mathrm{var}\, \varepsilon(t) = 1$. Hence

$$C(q^{-1})C(q) = (1+cq^{-1})(1+cq) = (c+q^{-1})(c+q) = c^2(1+c^{-1}q^{-1})(1+c^{-1}q).$$

The stochastic process $\{v(t)\}$ can thus also be represented as

$$v(t) = (1 + \frac{1}{c}q^{-1}) c\varepsilon(t) = \varepsilon'(t) + \frac{1}{c} \varepsilon'(t-1),$$

where $\mathrm{var}\, \varepsilon'(t) = c^2$.

□

Multivariable Generalizations

The CARMA model can easily be generalized to the multivariable case. The description (2.4) still holds provided that $y(t)$, $u(t)$, and $\varepsilon(t)$ are interpreted as vectors and that $A(q^{-1})$, $B(q^{-1})$, and $C(q^{-1})$ are interpreted as matrix polynomials. The vectors y and ε can be chosen to be of the same dimension. The matrix polynomial $C(q^{-1})$ can always be chosen in such a way that $\det C(\xi)$ will always have its zeros outside the unit disc or on the unit circle.

The multivariable CARMA model can be used to represent input output relations for multivariable industrial regulation problems. This model is also used in economics to represent the so called reduced form of a macro economic model.

3. OPTIMAL PREDICTION

The optimal prediction problem will now be discussed as a preliminary to solve the minimum variance control problem. The main result is given by

THEOREM 3.1
Let $\{y(t), t = 0, \pm 1, \pm 2, \ldots\}$ be a normal stochastic process with the representation

$$A(q^{-1}) y(t) = C(q^{-1}) \varepsilon(t), \tag{3.1}$$

where $\{\varepsilon(t), t = 0, \pm 1, \pm 2, \ldots\}$ is a sequence of independent normal $(0, R)$ random variables. Assume that the polynomial $\det C(\xi)$ has all its zeros outside the unit disc. Then the k-step predictor which minimizes the variance of the prediction error in steady state is given by

$$\hat{y}(t+k|t) = G(q^{-1}) C^{-1}(q^{-1}) y(t), \tag{3.2}$$

where

$$A^{-1}(q^{-1})C(q^{-1}) = C(q^{-1})A^{-1}(q^{-1}) = F(q^{-1}) + q^{-k}G(q^{-1})A^{-1}(q^{-1}) \quad (3.3)$$

and the polynomial $F(q^{-1})$ is of degree k-1:

$$F(q^{-1}) = I + F_1 q^{-1} + \ldots + F_{k-1} q^{-k+1}. \quad (3.4)$$

The error of the optimal predictor is a moving average of order k

$$\tilde{y}(t+k|t) = \varepsilon(t+k) + F_1 \varepsilon(t+k-1) + \ldots + F_{k-1} \varepsilon(t+1) \quad (3.5)$$

and the covariance of the prediction error is

$$\text{cov}[\tilde{y},\tilde{y}] = R + F_1 R F_1^T + \ldots + F_{k-1} R F_{k-1}^T. \quad (3.6)$$

Proof:
The proof is straightforward and constructive. Equations (3.1) and (3.3) give

$$y(t+k) = CA^{-1}\varepsilon(t+k) = F\varepsilon(t+k) + GA^{-1}\varepsilon(t).$$

Substitution of ε by y in the last term using (3.1) gives

$$y(t+k) = F\varepsilon(t+k) + GC^{-1}y(t).$$

The expression $GC^{-1}y(t)$ exist because det C = det C and it was assumed that all zeros of det $C(\xi)$ were outside the unit disc. Now let $\hat{y}$ be an arbitrary function of $y(t), y(t-1), \ldots$ Consider the prediction error

$$\tilde{y}(t+k|t) = y(t+k) - \hat{y} = F\varepsilon(t+k) + [GC^{-1}y(t) - \hat{y}]. \quad (3.7)$$

Let a be an arbitrary vector. Then

$$E[a^T\tilde{y}(t+k|t)]^2 = E[a^T F\varepsilon(t+k)]^2 + E\{a^T[GC^{-1}y(t) - \hat{y}]\}^2 +$$
$$+ 2E\{a^T F\varepsilon(t+k) a^T[GC^{-1}y(t) - \hat{y}]\}. \quad (3.8)$$

The last term vanishes because $\varepsilon(t+k), \varepsilon(t+k-1), \ldots \varepsilon(t+1)$ are all independent of $y(t), y(t-1), \ldots$ and then also independent of $\hat{y}$. The predictor (3.2) thus gives the minimum value of the prediction error for all a. It then follows from (3.7) that the prediction error is given by (3.5). A simple calculation based on $\{\varepsilon(t)\}$ being independent then gives (3.6).

□

Remark 1
Notice that the best predictor is linear. The linearity does not

depend critically on the minimum variance criterium. Since $\{y(t)\}$ is normal, the result would be the same for all criteria of the form $E\ h\{a^T[y(t+k) - \hat{y}]\}$ provided that h is symmetrical.

Remark 2
The assumption that $\varepsilon(t)$ and $\varepsilon(s)$ are independent for $t \neq s$ is crucial for the argument that the last term in (3.8) will vanish. If the stochastic variables $\varepsilon(t)$ and $\varepsilon(s)$ are not independent it is in general not true that the product of $\varepsilon(t+\tau)$ and an arbitrary function of $y(t), y(t-1), \ldots$ will vanish. However if the predictor is restricted to be linear functions of $y(t), y(t-1), \ldots$ then it is sufficient to assume $\varepsilon(t)$ and $\varepsilon(s)$ uncorrelated for the proof to hold. This situation is typical for linear problems with quadratic criteria.

Remark 3
Notice that it follows from (3.5) that

$$\tilde{y}(t+1|t) = y(t+1) - \hat{y}(t+1|t) = \varepsilon(t+1).$$

The stochastic variables $\{\varepsilon(t)\}$ can thus be interpreted as the innovations of the stochastic process $\{y(t)\}$. It is straightforward to calculate the predictor. The polynomials A and C such that $CA^{-1} = A^{-1}C$ are first determined. The polynomials F and G are then obtained as the quotient of degree $k-1$ and the remainder obtained when dividing C by A.

Notice that the predictor is a dynamical system (3.2) whose dynamics is governed by the matrix polynomial $C(\xi)$. The assumption in the theorem that $\det C(\xi)$ has all its zeros outside the unit disc guarantees that the predictor is stable. The initial conditions chosen for the predictor are thus immaterial. It was shown in Section 2 that the model could always be chosen in such a way that $\det C(\xi)$ has all its zeros outside the unit disc or on the unit circle. The Theorem 3.1 thus assumes away the case when $\det C(\xi)$ has zeros on the unit circle. This case requires special treatment because the optimal predictor is timevarying. A simple example illustrates what happens.

EXAMPLE 3.1
Consider the following scalar process

$$y(t) = \varepsilon(t) - \varepsilon(t-1).$$

In this case the polynomial $C(\xi) = 1 - \xi$ has apparently a zero on the unit circle. The one-step predictor is given by

$$\hat{y}(t+1|t) = -\varepsilon(t).$$

Attempting to calculate $\varepsilon(t)$ from $y(t), y(t-1), \ldots$ as was done before we get

$$\varepsilon(t) = \sum_{k=t_0+1}^{t} y(k) + \varepsilon(t_0) = z(t) + \varepsilon(t_0).$$

The presence of the term $\varepsilon(t_0)$ whose influence does not vanish as $t_0 \to -\infty$ clearly shows the consequences of the equation

$$C(q^{-1}) \varepsilon(t) = y(t)$$

being unstable. The initial condition $\varepsilon(t_0)$ can be estimated by

$$\hat{\varepsilon}(t_0) = -\frac{1}{t-t_0} \sum_{k=t_0+1}^{t} z(k) = -\frac{1}{t-t_0} \sum_{k=t_0+1}^{t} \sum_{i=t_0+1}^{k} y(i) =$$

$$= -\frac{1}{t-t_0} \sum_{k=t_0+1}^{t} (t+1-k) y(k).$$

This estimate will converge to $\varepsilon(t_0)$ as the number of terms in the series increases towards infinity. The predictor for y then becomes

$$\hat{y}(t+1|t) = -\sum_{k=t_0+1}^{t} \frac{k-1-t_0}{t-t_0} y(k) = -\sum_{i=1}^{t-t_0} \frac{t-t_0-i}{t-t_0} y(t+1-i). \tag{3.9}$$

This predictor is clearly not a linear time-invariant system. Notice that the predictor (3.9) has a variance that approaches $E\varepsilon^2$ as $t-t_0 \to \infty$. A formal application of Theorem 3.1 gives the predictor

$$\hat{y}(t+1|t) = -\sum_{-\infty}^{t} y(k),$$

which gives a prediction error with variance $2E\varepsilon^2$. □

The result of this example can be extended to the general case. See Hannan [5].

4. MINIMUM VARIANCE CONTROL

Having solved the prediction problem for the ARMA process we will now return to the CARMA process defined by equation (2.4) where A, B, and C are now regarded as matrix polynomials. To formulate the control problem it is necessary to define a criterion and the admissible controls.

The Criterion

It is assumed that the criterion for the control problem is to control the system in such a way that the steady state variance of the output is as small as possible. This criterion is a fairly good model for steady state control of important quality variables in industrial processes. The situation is illustrated in Fig. 2. Because of the fluctuations in the process output it is necessary to choose the reference value for the regulator above the test limit to make sure that a given percentage of the production is acceptable. By reducing the variance in the output it is then possible to operate closer to the test limit. This gives a gain which can be capitalized as increased production or reduction of raw materials used. For processes with a large volume of production even very moderate reductions in variance can give very substantial benefits.

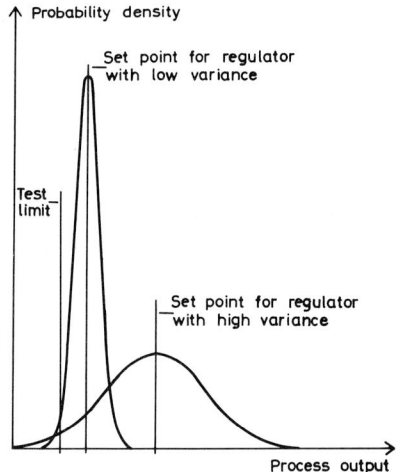

Fig. 2 - Illustrates that a decrease of the variance of the output signal makes it possible to move the set point closer to the test limit

For single output systems the criterion will thus be taken as to minimize the following loss function

$$V_1 = E\, y^2(t). \tag{4.1}$$

It will be shown that the same results will be obtained for the loss-function

$$V_\infty = \lim_{N\to\infty} E \frac{1}{N} \sum_{t=1}^{N} y^2(t). \qquad (4.2)$$

In the vector case the corresponding lossfunctions are

$$V_1 = E\, y^T(t)\, Q\, y(t) \qquad (4.3)$$

and

$$V_\infty = \lim_{N\to\infty} \frac{1}{N} E \sum_{t=1}^{N} y^T(t)\, Q\, y(t) \qquad (4.4)$$

respectively.

Admissible Controls

It is assumed that the admissible control laws are such that $u(t)$, i.e. the value of the control signal at time t, is a function of $y(t), y(t-1),\ldots$ and $u(t-1), u(t-2),\ldots$. By restricting the function to be linear the assumption on $\{\varepsilon(t)\}$ in the CARMA model can be relaxed from $\varepsilon(t)$ and $\varepsilon(s)$ being independent for $t \neq s$ to being uncorrelated.

The Minimum Variance Control Problem

The problem of controlling a CARMA process in such a way that the minimum variance criterion (4.1) or (4.3) is minimized will now be discussed. The solution is given by

THEOREM 4.1
Consider a CARMA process given by (2.4) where $\{\varepsilon(t)\}$ is a sequence of independent stochastic vectors with zero means and covariances R. Assume that the number of inputs and outputs are the same and that the polynomials $\det C(\xi)$ and $\det B(\xi)$ have all their zeros outside the unit disc. Let the matrix polynomials $F(q^{-1})$ of degree k-1 and $G(q^{-1})$ of degree n-1 be defined by

$$A^{-1}(q^{-1})\, C(q^{-1}) = F(q^{-1}) + q^{-k} A^{-1}(q^{-1})\, G(q^{-1}). \qquad (4.5)$$

Then the control law

$$u(t) = -B^{-1}(q^{-1})\, G(q^{-1})\, F^{-1}(q^{-1})\, y(t) = -B^{-1}(q^{-1})\, G(q^{-1})\, \varepsilon(t) \qquad (4.6)$$

minimizes the criterion (4.3) in the steady state and the steady output of the controlled system becomes

$$y(t) = F(q^{-1})\, \varepsilon(t) = \varepsilon(t) + F_1 \varepsilon(t-1) + \ldots + F_{k-1} \varepsilon(t-k+1). \qquad (4.7)$$

Proof:

A change of the control signal at time t will be noticeable in the output at first at time $t+k$. Because the matrix B_0 was assumed regular it is also possible to change all components of the output at time $t+k$ arbitrarily. It follows from (2.4) and (4.5) that

$$y(t+k) = F(q^{-1}) \varepsilon(t+k) + A^{-1}(q^{-1})[B(q^{-1})u(t) + G(q^{-1})\varepsilon(t)].$$

For simplicity the polynomial $A(q^{-1})$ will now simply be written as A. Using (2.4) to eliminate ε in the last term we get

$$y(t+k) = F\varepsilon(t+k) + A^{-1}Bu(t) + A^{-1}GC^{-1}Ay(t) - A^{-1}GC^{-1}Bu(t-k) =$$

$$= F\varepsilon(t+k) + FC^{-1}Bu(t) + A^{-1}GC^{-1}Ay(t), \qquad (4.8)$$

where the equality is obtained by applying (4.5) to the terms containing $u(t)$. To proceed notice that

$$G(AF)^{-1} C = C(AF)^{-1} G, \qquad (4.9)$$

because it follows from (4.5) that

$$G(AF)^{-1} C = q^k(C-AF)(AF)^{-1}C = q^k[C(AF)^{-1}C - C]$$

$$C(AF)^{-1} G = q^k C(AF)^{-1}(C-AF) = q^k[C(AF)^{-1}C - C].$$

Equations (4.8) and (4.9) give

$$y(t+k) = F(q^{-1})\varepsilon(t+k) + F(q^{-1})C^{-1}(q^{-1})[B(q^{-1})u(t) + G(q^{-1})F^{-1}(q^{-1})y(t)].$$

The two terms of the right member are independent because of the definition of admissible strategies, because the polynomial $\det C(q^{-1})$ is stable too, and because of $\varepsilon(t+k)$ being independent of $y(t), y(t-1),$ ... for $k > 0$. It thus follows that

$$E\, y^T(t+k)Q\,y(t+k) \geq E[F(q^{-1})\varepsilon(t+k)]^T Q[F(q^{-1})\varepsilon(t+k)] =$$

$$= \mathrm{tr}\,[Q + F_1^T Q F_1 + \ldots + F_{k-1}^T Q F_{k-1}]\,R,$$

where equality is obtained for

$$B(q^{-1})u(t) + G(q^{-1})\varepsilon(t) = 0.$$

Then also

$$y(t) = F(q^{-1})\varepsilon(t).$$

A combination of these equations gives the control law (4.6). To see the transient behaviour of the system introduce the control law (4.6) into the system description (2.4). Hence

$$[A(q^{-1}) + q^{-k}G(q^{-1})F^{-1}(q^{-1})] y(t) = C(q^{-1}) \varepsilon(t).$$

Equation (4.5) gives

$$C(q^{-1}) [F^{-1}(q^{-1})y(t) - \varepsilon(t)] = 0. \tag{4.10}$$

Since the polynomial $C(\xi)$ was assumed to have all its zeros outside the unit disc, this implies that the expression in brackets will converge to zero exponentially at a rate governed by the zeros of det $C(\xi)$.

Remark 1
The theorem still holds if $\varepsilon(t)$ and $\varepsilon(s)$ are only assumed uncorrelated for $t \neq s$ if a linear control law is postulated.

Remark 2
A comparison with the solution of the prediction problem shows that the control error under minimum variance control equals the error in predicting the process k steps. The minimum variance control law can thus be interpreted as doing the following. Predict output k steps ahead where k is the time it takes before a control action is noticeable in the output. Choose a control signal which makes the predicted value equal to the desired output.

Remark 3
The control error is a moving average of order k. This is easy to test and useful for diagnosis.

Remark 4
It follows from (4.10) that the poles of the closed loop system are given by

$$\det C(z^{-1}) = 0.$$

Remark 5
Notice that the control law (4.6) does not depend on the matrix Q. The control law will thus simultaneously minimize the variances in all components of the output. This motivates the name minimum variance strategy.

Control Effectiveness

Without control the output becomes

$$y_0(t) = A^{-1}(q^{-1}) C(q^{-1}) \varepsilon(t)$$

and under minimum variance control the output becomes

$$y_{mv}(t) = F(q^{-1}) \varepsilon(t).$$

Hence

$$y_{mv}(t) = F(q^{-1}) C^{-1}(q^{-1}) A(q^{-1}) y_0(t).$$

The reduction of the fluctuations in the output can thus be characterized by the transfer function

$$H(q^{-1}) = F(q^{-1}) C^{-1}(q^{-1}) A(q^{-1}).$$

A simple example illustrates what can happen.

EXAMPLE

Consider a first order scalar system with $k = 1$ and

$$A(q^{-1}) = 1 + a q^{-1}$$

$$C(q^{-1}) = 1 + c q^{-1}.$$

Hence

$$H(q^{-1}) = \frac{1 + aq^{-1}}{1 + cq^{-1}}$$

$$|H(e^{-i\omega})| = \frac{1 + a^2 + 2a \cos \omega}{1 + c^2 + 2c \cos \omega}.$$

A graph of the function $|H|$ is shown in Fig. 3. The graph shows that in the particular case the action of the minimum variance is to reduce the low-frequency components and to increase the high frequency components in the output.

The Minimum Phase Assumption

In Theorem 4.2 it was assumed that the polynomial $\det B(\xi)$ has all its zeros outside the unit disc. This assumption is called the minimum phase condition because it implies that the input-output relation given by (2.4) for $\varepsilon = 0$ is a nonminimum phase system. If this condition is violated the control law given by (4.6) still gives the

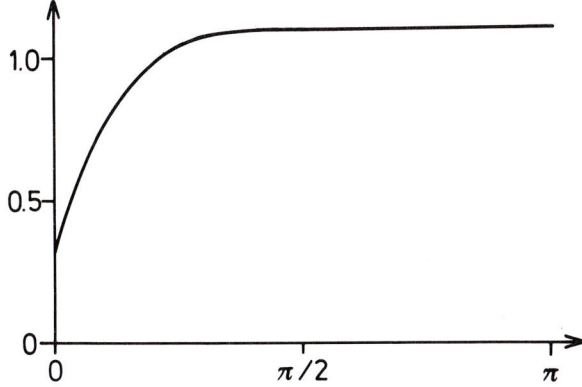

Fig. 3 - Amplitude curve of the transfer function H which shows how the minimum variance control law attenuates disturbances of different frequencies

smallest variance of the outputs. It follows, however, from (4.6) that the control signal u is given by

$$u(t) = -B^{-1}(q^{-1}) \, G(q^{-1}) \, \varepsilon(t).$$

If the polynomial $\det B(\xi)$ has zeros inside the unit disc this difference equation will be unstable and the control signal u will grow exponentially. This will not have any influence on the output y because the exponential components of u will be cancelled by the operator $B(q^{-1})$ which operates on y in the system model. The cancellation will of course only be possible if the control law was calculated from a precise model of the system. Small perturbations in the model implies that the exponentially growing components will be transmitted to the output. This is illustrated in Fig. 4 which shows the results of a simulation.

From a practical point of view it is thus clear that the control law (4.6) is useless if the polynomial $\det B(\xi)$ has zeros inside the unit disc. There are several different possibilities to circumvent this problem. One possibility is to include a penalty on the control actions, i.e. to change the criterion to

$$E[y^2 + \rho u^2].$$

Control laws with the property that u(t) will be very large will then be excluded. In several cases it may, however, be unrealistic to assign a proper value of ρ and we will therefore investigate the problem with $\rho = 0$. It turns out in fact that the problem of minimizing (4.1) has several local minima if the polynomial $\det B(\xi)$ has

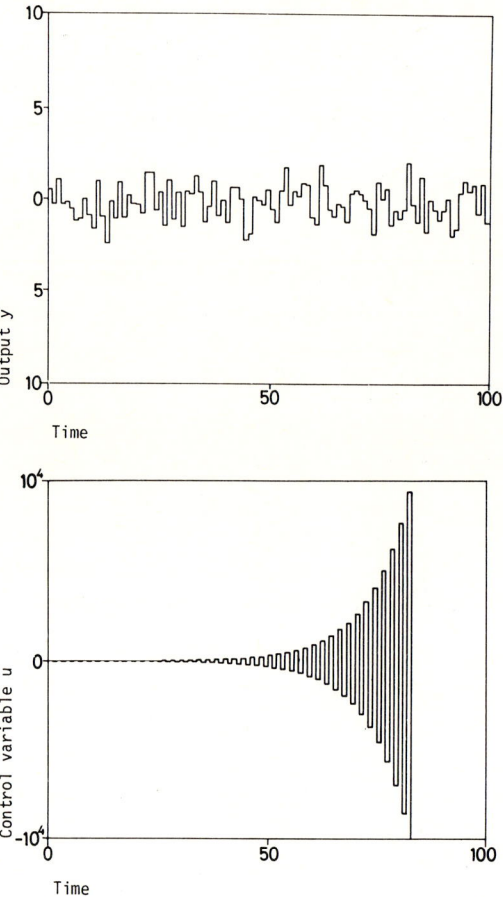

Fig. 4 - Simulation of a nonminimum phase system with minimum variance control. The system is described by
$y(t) - 1.7\, y(t-1) + 0.7\, y(t-2) = 0.9\, u(t-1) + u(t-2) + e(t) - 0.7\, e(t-1)$

zeros inside the unit disc. Before stating the main result the notion of reciprocal polynomial is introduced. Let $B(\xi)$ be a polynomial

$$B(\xi) = b_0 + b_1 \xi + \ldots + b_n \xi^n \; ;$$

then the reciprocal polynomial is defined by

$$\tilde{B}(\xi) = b_0 \xi^n + b_1 \xi^{n-1} + \ldots + b_n .$$

In the single-input single-output case we have

THEOREM 4.2

Consider a single-input single-output system described by (2.4). Assume that the polynomial $C(\xi)$ has all its zeros outside the unit disc and that the polynomial $B(\xi)$ has zeros both inside and outside the unit disc. Let B be factored as

$$B(q^{-1}) = B_1(q^{-1}) B_2(q^{-1}), \qquad (4.11)$$

where $B_2(\xi)$ has all its zeros inside the unit disc and $B_2(0) = 1$. Let $H(q^{-1})$ and $K(q^{-1})$ be defined by the partial fraction expansion

$$\frac{C(q^{-1}) \tilde{B}_2(q^{-1})}{A(q^{-1}) q^{-k} B_2(q^{-1})} = \frac{H(q^{-1})}{q^{-k} B_2(q^{-1})} + \frac{K(q^{-1})}{A(q^{-1})}, \qquad (4.12)$$

where

$$\deg H(q^{-1}) = k - 1 + \deg B_2(q^{-1}). \qquad (4.13)$$

Then the variance of the output has a local minimum for the control law

$$u(t) = -\frac{K(q^{-1})}{H(q^{-1}) B_1(q^{-1})} y(t) = -\frac{K(q^{-1})}{B_1(q^{-1}) \tilde{B}_2(q^{-1})} \varepsilon(t) \qquad (4.14)$$

and the corresponding output is given by

$$y(t) = \frac{H(q^{-1})}{\tilde{B}_2(q^{-1})} \varepsilon(t). \qquad (4.15)$$

Proof:
Equation (2.4) gives

$$y(t+k) = \frac{B}{A} u(t) + \frac{C}{A} \varepsilon(t+k) = \frac{B_2}{\tilde{B}_2} w(t+k),$$

where

$$w(t+k) = \frac{B_1 \tilde{B}_2}{A} u(t) + \frac{C \tilde{B}_2}{q^{-k} A B_2} \varepsilon(t).$$

The signals y and w have the same variances because B_2 and $\tilde{B}_2$ are reciprocal polynomials. Equation (4.12) then gives

$$w(t+k) = \frac{B_1 \tilde{B}_2}{A} u(t) + \frac{H}{q^{-k} B_2} \varepsilon(t) + \frac{K}{A} \varepsilon(t).$$

Now use (2.4) to eliminate $\varepsilon(t)$ in the last term, then

$$w(t+k) = \frac{H}{q^{-k}B_2}\varepsilon(t) + \frac{B_1\tilde{B}_2}{A}u(t) + \frac{K}{A}\left[\frac{A}{C}y(t) - \frac{B}{C}u(t-k)\right] =$$

$$= \frac{H}{q^{-k}B_2}\varepsilon(t) + \left[\frac{B_1\tilde{B}_2}{A} - \frac{q^{-k}KB}{AC}\right]u(t) + \frac{K}{C}y(t) =$$

$$= \frac{H}{q^{-k}B_2}\varepsilon(t) + \frac{q^{-k}B_1B_2}{C}\left[\frac{C\tilde{B}_2}{Aq^{-k}B_2} - \frac{K}{A}\right]u(t) + \frac{K}{C}y(t) =$$

$$= \frac{H}{q^{-k}B_2}\varepsilon(t) + \frac{B_1H}{C}u(t) + \frac{K}{C}y(t),$$

where the last equality follows from (4.12). Hence

$$w(t+k) = \frac{H(q^{-1})}{B_2(q^{-1})}\varepsilon(t+k) + \left[\frac{B_1(q^{-1})H(q^{-1})}{C(q^{-1})}u(t) + \frac{K(q^{-1})}{C(q^{-1})}y(t)\right]. \quad (4.16)$$

Because the polynomial $B_2(\xi)$ has all its zeros inside the unit disc and because of (4.13) the first term of the right member of (4.16) can be written as the converging series

$$\frac{H(q^{-1})}{q^{-k}B_2(q^{-1})}\varepsilon(t) = \varepsilon(t+1) + \alpha_1\varepsilon(t+2) + \dots \quad .$$

Since the polynomial $C(\xi)$ has all its zeros outside the unit disc, the second term can be expanded as a converging series in $y(t), y(t-1),\dots u(t), u(t-1),\dots$ The two terms of (4.16) are thus independent and the smallest variance is obtained for the control law (4.14). The output of the controlled system is then given by (4.15).

Remark

The control signal u defined by (4.14) is bounded if $B_2(\xi)$ is chosen as the factor which contains all zeros of B inside the unit disc.

Consider a system described by (2.4) where the polynomial $B(\xi)$ has zeros inside the unit disc. According to Theorem 4.1 there is an absolute minimum to the variance of the output given by

$$E[F(q^{-1})\varepsilon(t)]^2. \quad (4.17)$$

The control law which realizes this minimum is given by

$$u(t) = -\frac{G(q^{-1})}{B(q^{-1})F(q^{-1})}y(t) = -\frac{G(q^{-1})}{B(q^{-1})}\varepsilon(t).$$

The variance of the control signal will clearly be infinite because $B(q^{-1})$ is unstable. Let B_2 be an unstable factor of B. According to Theorem 4.2 there is then another local minimum of the loss function given by

$$E\left[\frac{H(q^{-1})}{q^{-k}\tilde{B}_2(q^{-1})}\varepsilon(t)\right]^2. \tag{4.18}$$

It is easy to show that

$$\frac{H(q^{-1})}{q^{-k}\tilde{B}_2(q^{-1})}\varepsilon(t) = F(q^{-1})\varepsilon(t+k) + \frac{L_1(q^{-1})}{\tilde{B}_2(q^{-1})}\varepsilon(t),$$

where L_1 is defined by the partial fraction expansion

$$\frac{G(q^{-1})B_2(q^{-1})}{A(q^{-1})B_2(q^{-1})} = \frac{L_1(q^{-1})}{B_2(q^{-1})} + \frac{L_2(q^{-1})}{A(q^{-1})} \qquad \deg L_1 < \deg B_2.$$

The loss function (4.18) is thus always larger than (4.17) and the term

$$E\left[\frac{L_1(q^{-1})}{\tilde{B}_2(q^{-1})}\varepsilon(t)\right]^2$$

represents the increase in the loss function required in order to avoid having the factor $B_2(q^{-1})$ in the equation (4.14).

When solving the minimum variance problem B_2 should be chosen to contain all factors of B which have zeros inside the unit disc except those factors which are also factors of A.

5. APPLICATIONS

The minimum variance control theory tells that the feedback law of Fig. 1 is a time invariant dynamical system. The theory also gives a possibility to interpret the action of the feedback law. In the simplest case action of the feedback can be described as follows: Predict the output k steps ahead. Choose a control signal such that the predicted value is equal to the desired value. The complexity of the feedback law is uniquely determined by the mathematical model of the process. For simple models the regulator may be equivalent to the common PI regulator (1) but the regulator may also be much more complicated.

The minimum variance regulator has been applied to a number of different industrial control problems. The problems have typically been steady state regulation of industrial processes. The benefits obtained have been determined from the arguments illustrated in Fig. 2. The successful applications have the property that moderate reductions of the variances will give rise to substantial economic gains. It is thus motivated to spend the extra effort required to develop the models and to obtain the control laws. The determination of the mathematical model (2.4) is the major difficulty when trying to apply minimum variance control. The model (2.4) can rarely be obtained from apriori physical data. Instead the model has to be estimated from data obtained from an experiment on the process. In a typical experiment the input signal is perturbed and the resulting variations in the output are recorded. When the model is obtained it is possible to tell the results that can be expected from minimum variance control. This is a substantial advantage because it can then be decided if the effort is worthwhile and if it is justified to use a control law which is more complicated than the simple PID regulator.

6. REFERENCES

The scalar version of the minimum variance control law is discussed in [1]. This reference also contains many references. The solution to the nonminimum phase problem was given in [2]. The spectral factorization idea is given in [3]. An application of the minimum variance strategy to paper machine control is described in [2]. Applications to metal rolling are described in [4].

[1]   K J Åström: Introduction to Stochastic Control Theory. Academic Press, New York 1970.

[2]   V Peterka: On Steady State Minimum Variance Control Strategy. Kybernetika $\underline{8}$ (1972) 219-232.

[3]   N Wiener: Extrapolation, Interpolation and Smoothing of Stationary Time Series with Engineering Applications. Wiley, New York 1949.

[4]   D Watanapongse and N A Robbins: Application of Modern Control Techniques to Computerized Setup for Effective Operation of Inland Strip Mills. Report, Inland Steel Company, East Chicago, Indiana, USA, 1976.

[5]   E J Hannan: Multiple Time Series. Wiley, New York 1970.

CHAPTER 3 - LINEAR QUADRATIC GAUSSIAN CONTROL

1. INTRODUCTION

The theory described in this chapter is called linear quadratic gaussian (LQG) control theory because the process dynamics is characterized by linear equations, the criterion is a quadratic function, and the disturbances are gaussian. In the previous chapter the process dynamics was also described by an external model. This model was a difference equation which related the process output to its input and the disturbances. In this chapter the process model will instead be described by an internal model. A set of variables which completely specifies the past development of the system called *state variables* are thus introduced. The mathematical model is then a difference equation which describes the future development of the state variables. The model is still assumed to be linear. The criterion is the expected value of a general quadratic form. The problem is thus slightly more general than the problem discussed in the previous chapter. The development of the theory is also analogous to that of the previous chapter. The mathematical models used are discussed in Section 2. The prediction problem is then solved in Section 3 and the control problem is solved in Section 4.

When external models are used it was natural to use the theory of polynomials and rational functions. For the internal models it is instead natural to use matrix theory. The relations beween the two approaches to the problem are discussed in Section 5 and Section 6 deals briefly with applications. Since the LQG theory is covered in detail in text books, the treatment is here kept fairly brief.

2. MATHEMATICAL MODELS

Process Dynamics and Disturbances

For simplicity only discrete time systems will be considered. It is assumed that time T is the set of integers $\{\ldots -1, 0, 1, \ldots\}$. Let the input u, the state x, and the output y be vector valued time functions of dimensions p, n, and r. It is assumed that the system and its environment can be described by the linear difference equations

$$x(t+1) = Ax(t) + Bu(t) + v(t)$$
$$y(t) = Cx(t) + e(t), \qquad t \in T, \qquad (2.1)$$

where the "process noise" $\{v(t), t \in T\}$ and the "measurement noise" $\{e(t), t \in T\}$ are sequences of independent gaussian random vectors. It is assumed that $\{v(t)\}$ and $\{e(t)\}$ have zero mean values and that their covariances are given by

$$\text{cov}[v(t), v(s)] = \begin{cases} R_1 & t = s \\ 0 & t \neq s \end{cases}$$

$$\text{cov } v(t), e(s) = 0$$

$$\text{cov } e(t), e(s) = \begin{cases} R_2 & t = s \\ 0 & t \neq s. \end{cases} \quad (2.2)$$

The initial condition $x(t_0)$ of (2.1) is assumed to be gaussian with mean value m and covariance R_0. It is also assumed that the initial condition is independent of $\{v(t)\}$ and $\{e(t)\}$ or equivalently that $x(t)$ is independent of $\{v(t)\}$ and $\{e(t)\}$.

The model (2.1) tells that given the values of the state x and the control u at time t then the conditional distribution of the state at time $t+1$ is gaussian with mean value $Ax(t) + Bu(t)$ and covariance R_1. The equation (2.1) also tells that the conditional distribution of the measurement $y(t)$ given $x(t)$ is gaussian with mean $Cx(t)$ and covariance R_2.

Notice that it is frequently necessary to introduce extra variables in order to arrive at a model having the form (2.1). For example if the environment is characterized by a disturbance having the spectral density

$$\phi(\omega) = \frac{1}{1 + a^2 - 2a \cos \omega}$$

it can be characterized by the difference equation

$$\xi(t+1) = a\xi(t) + n(t),$$

where $\{n(t)\}$ is white noise. It is then necessary to include ξ as a component of the state vector. Similarly a constant disturbance acting on the system can be described by the difference equation

$$d(t+1) = d(t).$$

Such a disturbance can be included by augmenting the state vector.

The Criterion

In the linear quadratic gaussian problem it is assumed that the purpose of the control can be expressed as to minimize the loss function

$$V_1 = \min E \left\{ x^T(N)Q_0 x(N) + \sum_{t=t_0}^{N-1} x^T(t)Q_1 x(t) + u^T(t)Q_2 u(t) \right\}. \qquad (2.3)$$

Time Varying Models

The matrices A, B, C, R_1, R_2, Q_1, and Q_2 may vary with time t.

3. KALMAN FILTERING AND PREDICTION

The filtering problem will be solved before the optimal control problem is discussed. It is assumed that the outputs $y(t_0),\ldots,y(t)$ have been observed and the problem is to predict $x(t+1)$ as well as possible. Let $\mathcal{Y}_t$ denote the σ-algebra generated by $y(t),\ldots,y(t_0)$. The prediction problem is clearly solved if the conditional distribution of $x(t+1)$ given $\mathcal{Y}_t$ can be determined. The solution is given by the following theorem.

THEOREM 3.1

Let the gaussian process $\{x(t)\}$ be generated by (2.1) with $u=0$. The conditional distribution of $\{x(t+1)\}$ given $\mathcal{Y}_t$ is gaussian $(\hat{x}(t+1), P(t+1))$ where

$$\begin{cases} \hat{x}(k+1) = A\hat{x}(k) + K(k)[y(k) - C\hat{x}(k)], & k = t_0,\ldots,t \\ \hat{x}(t_0) = m \end{cases} \qquad (3.1)$$

$$K(k) = AP(k)C^T[CP(k)C^T + R_2]^{-1} \qquad (3.2)$$

$$\begin{cases} P(k+1) = AP(k)A^T + R_1 - AP(k)C^T[CP(k)C^T + R_2]^{-1}CP(k)A^T = \\ \qquad\quad = [A - K(k)C]P(k)A^T + R_1, \quad k = t_0,\ldots,t \\ P(t_0) = R_0. \end{cases} \qquad (3.3)$$

Proof:
The proof consists of a repeated use of the following well known property of gaussian random variables. If the vector

$$\begin{pmatrix} x \\ y \end{pmatrix}$$

is gaussian with mean value

$$\begin{pmatrix} m_x \\ m_y \end{pmatrix}$$

and covariance

$$\begin{pmatrix} R_x & R_{xy} \\ R_{yx} & R_y \end{pmatrix},$$

then the conditional mean of x given y is

$$E[x|y] = m_x + R_{xy} R_y^{-1} (y - m_y).$$

Full details are given in the references.

□

Remark 1

The theorem has a strong intuitive appeal. The term $A\hat{x}(k)$ is the apriori estimate of $x(k+1)$ and the correction to the prior $K(k)[y(k) - C\hat{x}(k)]$ is proportional to the deviation of the measurement $y(k)$ from its prior $C\hat{x}(k)$.

Remark 2

The covariance $P(k)$ does not depend on the measurements.

Remark 3

The result of the theorem can easily be extended to include a control signal different from zero in (2.1). If $u(t)$ is measurable with respect to y_t for each t then it is easily shown that the conditional distribution of $x(t+1)$ given y_t is gaussian $(\hat{x}(t+1), P(t))$ where

$$\hat{x}(t+1) = A\hat{x}(t) + Bu(t) + K(t)[y(t) - C\hat{x}(t)] \tag{3.4}$$

and $K(t)$ and $P(t)$ are given by (3.2) and (3.3).

Remark 4

The theorem can be extended to the case when the random processes $\{v(t)\}$ and $\{e(t)\}$ are assumed to be second order processes only. The best *linear* prediction is then given by $\hat{x}(t+1)$.

Innovations Representations

Theorem 3.1 allows for an alternative representation of the stochastic process $\{y(t)\}$. It follows from the proof of Theorem 3.1 that the variables

$$\tilde{y}(t) = y(t) - C\hat{x}(t)$$

are gaussian random variables with zero mean values and the covariances

$$E \, \tilde{y}(t) \, \tilde{y}^T(s) = \begin{cases} R = [CP(t)C^T + R_2] & t = s \\ 0 & t \neq s. \end{cases} \qquad (3.5)$$

Since $\tilde{y}(t)$ is gaussian it then follows that $\{\tilde{y}(t), t \in T\}$ is a sequence of independent gaussian random variables. The following theorem is then obtained.

THEOREM 3.2
Consider the stochastic process $\{y(t)\}$ defined by (2.1) where $u(t)$ is measurable with respect to Y_t. The process $\{y(t)\}$ then has the representation

$$\begin{cases} \hat{x}(t+1) = A\hat{x}(t) + Bu(t) + K(t)\,\tilde{y}(t) \\ y(t) = C\hat{x}(t) + \tilde{y}(t), \end{cases} \qquad (3.6)$$

where $\{\tilde{y}(t)\}$ is a sequence of independent gaussian $(0, R)$ random variables where $K(t)$ is given by (3.2) and R by (3.5).

Duality

Let x and y be gaussian random vectors. The space obtained by introducing the scalar product

$$<x, y> = E \, x^T y$$

can be shown to be the dual of a Euclidean space. By using this concept of duality it can be shown that the Kalman filtering problem is the dual of a deterministic control problem.
To see this consider the problem of estimating $a^T x(t_1)$ linearly in $y(t_1-1), \ldots, y(t_0)$ and m in such a way that the criterion

$$E[a^T x(t_1) - a^T \hat{x}(t_1)]^2 \qquad (3.7)$$

is minimal.
As the estimate is linear we have

$$a^T \hat{x}(t_1) = - \sum_{t=t_0}^{t_1-1} u^T(t) \, y(t) + b^T m. \qquad (3.8)$$

The minus sign is introduced in order to obtain the final result in a nice form. The estimation problem is thus a problem of determining the vectors $b, u(t_1-1), u(t_1-2), \ldots, u(t_0)$. Now determine the u:s in such a way that the criterion (3.7) is minimal. To do so, introduce the vectors $z(t)$ defined recursively from

$$z(t) = A^T z(t+1) + C^T u(t+1) \tag{3.9}$$

with the initial condition

$z(t_1-1) = a.$

Hence

$$a^T x(t_1) = z^T(t_1-1)x(t_1) = z^T(t_0-1)x(t_0) + \sum_{t=t_0}^{t_1-1} [z^T(t)x(t+1) - z^T(t-1)x(t)]. \tag{3.10}$$

It follows from (2.1) and (3.9) that

$$z^T(t)x(t+1) = z^T(t)Ax(t) + z^T(t)v(t)$$
$$z^T(t-1)x(t) = z^T(t)Ax(t) + u^T(t)Cx(t).$$

Introducing this in (3.10), we find

$$a^T x(t_1) = z^T(t_0-1)x(t_0) + \sum_{t=t_0}^{t_1-1} [z^T(t)v(t) - u^T(t)Cx(t)]. \tag{3.11}$$

Equations (2.1) and (3.8) give

$$a^T \hat{x}(t_1) = -\sum_{t=t_0}^{t_1-1} u^T(t)y(t) + b^T m = -\sum_{t=t_0}^{t_1-1} [u^T(t)Cx(t) + u^T(t)e(t)] + b^T m. \tag{3.12}$$

Hence

$$a^T x(t_1) - a^T \hat{x}(t_1) = z^T(t_0-1)x(t_0) - b^T m + \sum_{t=t_0}^{t_1-1} [z^T(t)v(t) - u^T(t)e(t)].$$

Squaring and taking mathematical expectations, the criterion (3.7) can be expressed as follows:

$$E[a^T x(t_1) - a^T \hat{x}(t_1)]^2 = [(z(t_0-1) - b)^T m]^2 + z^T(t_0-1)R_0 z(t_0-1) +$$
$$+ \sum_{t=t_0}^{t_1-1} [z^T(t)R_1 z(t) + u^T(t)R_2 u(t)]. \tag{3.13}$$

To minimize the criterion, the parameter b must be chosen equal to $z(t_0-1)$ and the u:s should be determined in such a way that the function

$$z^T(t_0-1)R_0 z(t_0-1) + \sum_{t=t_0}^{t_1-1} [z^T(t)R_1 z(t) + u^T(t)R_2 u(t)] \tag{3.14}$$

is as small as possible.

It has now been shown that the problem of finding a linear predictor which minimizes (3.7) is equivalent to finding a control signal u for the system (3.9) such that the criterion (3.14) is minimal.

4. OPTIMAL CONTROL

Having solved the prediction problem we will now return to the optimal control problem. A system described by (2.1) is considered. The problem is to find an admissible control such that the criterion (2.3) is minimal. The following result is useful in the solution of the problem.

LEMMA 4.1

Consider a system described by the difference equation

$$x(t+1) = Ax(t) + Bu(t) + v(t). \tag{4.1}$$

Assume that the difference equation

$$S(t) = A^T S(t+1) A + Q_1 - A^T S(t+1) B [Q_2 + B^T S(t+1) B]^{-1} B^T S(t+1) A \tag{4.2}$$

with the initial condition

$$S(N) = Q_0 \tag{4.3}$$

has a solution $S(t)$ which is non-negative definite for $t_0 \leq t \leq N$ and such that

$$Q(t) = Q_2 + B^T S(t+1) B \tag{4.4}$$

is non-singular for all t. Let

$$L(t) = [Q_2 + B^T S(t+1) B]^{-1} B^T S(t+1) A. \tag{4.5}$$

Then

$$x^T(N) Q_0 x(N) + \sum_{t=t_0}^{N-1} x^T(t) Q_1 x(t) + u^T(t) Q_2 u(t) = x^T(t_0) S(t_0) x(t_0) +$$

$$+ \sum_{t=t_0}^{N-1} [u(t) + L(t) x(t)]^T [B^T S(t+1) B + Q_2] [u(t) + L(t) x(t)] +$$

$$+ \sum_{t=t_0}^{N-1} \left\{ v^T(t) S(t+1) [Ax(t) + Bu(t)] + \right.$$

$$\left. + [Ax(t) + Bu(t)]^T S(t+1) v(t) + v^T(t) S(t+1) v(t) \right\}. \tag{4.6}$$

Proof:

The proof is straightforward. We have the following

$$x^T(N)Q_0 x(N) = x^T(N)S(N)x(N) = x^T(t_0)S(t_0)x(t_0) +$$
$$+ \sum_{t=t_0}^{N-1} [x^T(t+1)S(t+1)x(t+1) - x^T(t)S(t)x(t)].$$

Consider the different terms of the sum. We have

$$x^T(t+1)S(t+1)x(t+1) = [Ax(t) + Bu(t) + v(t)]^T S(t+1)[Ax(t) + Bu(t) + v(t)]$$

and

$$x^T(t)S(t)x(t) = x^T(t)\{A^T S(t+1)A + Q_1 - L^T(t)[B^T S(t+1)B + Q_2]L(t)\}x(t).$$

Hence

$$x^T(N)Q_0 x(N) = x^T(t_0)S(t_0)x(t_0) + \sum_{t=t_0}^{N-1} \{[Ax(t) + Bu(t)]^T S(t+1)v(t) +$$
$$+ v^T(t)S(t+1)[Ax(t) + Bu(t)] + v^T(t)S(t+1)v(t)\} +$$
$$+ \sum_{t=t_0}^{N-1} \{u^T(t)[B^T S(t+1)B + Q_2]u(t) +$$
$$+ u^T(t)B^T S(t+1)Ax(t) + x^T(t)A^T S(t+1)Bu(t) +$$
$$+ x^T(t)L^T(t)[B^T S(t+1)B + Q_2]L(t)x(t) - x^T(t)Q_1 x(t) -$$
$$- u^T(t)Q_2 u(t)\},$$

where the term $u^T Q_2 u$ has been added and subtracted in the last sum. Rearrangement of the terms now completes the proof of the lemma. □

The Lemma 4.1 is a useful tool for solving the optimal control problem because it shows directly how the loss function is influenced by the value of the control signal at time t. The optimal control problem will now be solved for some different choices of the admissible controls.

Complete State Information

It is first assumed that the admissible controls are such that $u(t)$ is a function of $x(t)$. The solution to the optimal control problem is then given by

THEOREM 4.1

Consider a system described by (2.1). Let the admissible controls be such that $u(t)$ is a function of $x(t)$. Assume that the equation (4.2) with initial conditions (4.3) has a non-negative solution such that $Q(t)$ defined by (4.4) is positive definite for $t_0 \leq t \leq N$. Then the criterion (2.3) is minimal for the control law

$$u(t) = -L(t)x(t), \qquad (4.7)$$

where L is given by (4.5). The minimal loss is

$$\min V = m^T S(t_0) m + \operatorname{tr} S(t_0) R_0 + \sum_{t=t_0}^{N-1} \operatorname{tr} S(t+1) R_1(t). \qquad (4.8)$$

Proof:
Let x be gaussian (m, R). Then

$$E x^T Q x = m^T Q m + E(x-m)^T Q(x-m) = m^T Q m + E \operatorname{tr}(x-m)^T Q(x-m) =$$

$$= m^T Q m + E \operatorname{tr} Q(x-m)(x-m)^T = m^T Q m + \operatorname{tr} QR.$$

It follows from Lemma 4.1 that

$$E\left[x^T(N) Q_0 x(N) + \sum_{t=t_0}^{N-1} x^T(t) Q_1 x(t) + u^T(t) Q_2 u(t) \right] =$$

$$= m^T S(t_0) m + \operatorname{tr} S(t_0) R_0 + \sum_{t=t_0}^{N-1} \operatorname{tr} S(t+1) R_1(t) +$$

$$+ \sum [u(t) + L(t)x(t)]^T Q(t) [u(t) + L(t)x(t)], \qquad (4.9)$$

because $v(t)$ is independent of $x(t)$ and $u(t)$.

Since $Q(t)$ was assumed to be positive definite the right hand side is minimal for the control law (4.7) and the proof is completed.
□

Incomplete State Information

The admissible controls are now assumed to be such that $u(t+1)$ is a function of $y(t), \ldots, y(t_0)$ or more precisely for each t $u(t+1)$ is assumed to be measureable with respect to the σ-algebra Y_t generated by $y(t), \ldots, y(t_0)$. To obtain the result in this case a measure selection theorem will be used.

Let x and y be random variables which take values in R^n and R^p. Let $\ell(x,y,u)$ be a loss function $\ell: R^{n+p+r} \to R$. We have

LEMMA 4.2

Let $E[\cdot|y]$ denote the conditional mean given y. Assume that

$$f(y,u) = E[\ell(x,y,u)|y]$$

has a unique minimum with respect to u, attained for $u = u^0(y)$. Then the minimum of $E\ell(x,y,u(y))$ with respect to all $u(y)$ which are measurable with respect to y is given by

$$\min E\ell(x,y,u(y)) = E_y f(y,u^0(y)).$$

Proof:
The proof is given in Åström [2], p. 261. □

The solution to the control problem with incomplete state information is now given by

THEOREM 4.2 (Separation Theorem)

Consider a system described by (2.1). Let the admissible controls be such that for each t $u(t+1)$ is measurable with respect to y_t. Assume that the equation (4.2) with initial conditions (4.3) has a non-negative solution such that $Q(t)$ given by (4.4) is positive definite for $t_0 \leq t \leq N$. Then the criterion (2.3) is minimized for the control law

$$u(t) = -L(t)\hat{x}(t) \qquad (4.10)$$

where L is given by (4.5) and $\hat{x}(t)$ is the conditional mean of $x(t)$ given y_{t-1}, given by the Kalman filter (3.4). The minimal loss is given by

$$\min E V = m^T S(t_0) m + \operatorname{tr} S(t_0) R_0 + \sum_{t=t_0}^{N-1} \operatorname{tr} S(t+1) R_1(t) +$$
$$+ \sum_{t=t_0}^{N-1} \operatorname{tr} P(t) L^T(t) Q(t) L(t). \qquad (4.11)$$

Proof:
It follows from Remark 3 of Theorem 3.1 that the conditional distribution of $x(t+1)$ given y_t is given by (3.4) where the conditional covariance does not depend on the control signal.

Proceed in the same way as for the proof of Theorem 4.1 to obtain equation (4.9). Use Lemma 4.2 to minimize the right hand side of (4.9). The minimum (4.11) is then obtained for the control law (4.10). □

Remark 1

Notice that the different terms in the minimal loss function all have a nice physical interpretation. The term $m^T S(t_0) m$ is the contribution due to the off set of the initial state. The term $\operatorname{tr} S(t_0) R_0$ is the contribution due to the initial uncertainty of the initial state. The term $\Sigma \operatorname{tr} S(t+1) R_1(t)$ depends on the process noise that is acting on the system and the term $\Sigma \operatorname{tr} P(t) L^T(t) Q(t) L(t)$ depends on the uncertainty in the state estimation. A calculation of the relative magnitudes of the different terms will give good information about the nature of the difficulties in solving the control problem.

Remark 2

Notice that the optimal control law is a linear feedback from the conditional mean. The linear feedback gain L is the same as for the problem of Theorem 4.1 with complete state information. This motivates the name *certainty equivalence theorem* which is sometimes given to Theorem 4.2.

Theorem 4.2 gives valuable insight into the nature of the optimal feedback. The feedback can be thought of as being composed of two parts. See Fig. 5. One part is a dynamical system (a Kalman filter) which generates the conditional mean of the state vector from the measured process outputs. The other part is a static linear system which simply generates the control as a linear function (4.10) of the estimated state variables. See Fig. 5. Notice that the matrix L(t) in (4.10) only depends on A, B, Q_1, and Q_0 and that it is independent of the stochastic elements of the model. The gain K of the Kalman filter depends on A, C, R_0, R_1, and R_2 but it is independent of the loss function. This motivates the name *separation theorem*, which expresses the fact that the control problem can be split up into two parts: a deterministic control problem to obtain L and a Kalman filtering problem to obtain K. Also notice that the conditional covariance does not depend on the measured data.

To use Theorem 4.1 or Theorem 4.2 it must be asserted that the equations (3.3) and (4.2) have solutions. The conditions Q_2 and R_2 being positive definite and the system (2.1) being completely reachable and completely observable are sufficient to ensure this.

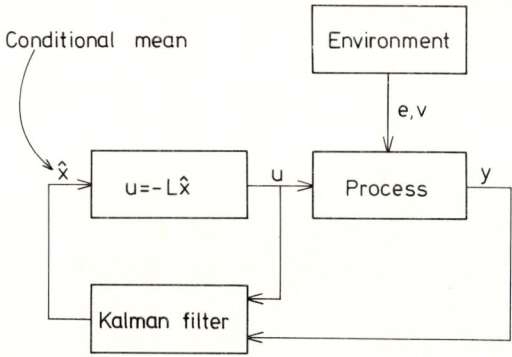

Fig. 5 - Block diagram which illustrates the feedback given by Theorem 4.2

Stationary Solutions

In many cases the matrices A, B, C, Q_1, Q_2, R_1, and R_2 which appear in the problem formulation are constant. Under weak additional assumptions it can then be shown that if $N \to \infty$ then the Kalman filter gain K and the feedback gain L will converge to unique constant solutions. A sufficient condition is that Q_2 and R_2 are positive and that the system (2.1) is completely reachable and completely observable. In such a case the optimal feedback shown in Fig. 5 is simply a linear time invariant dynamical system. There are, however, cases where the stationary solutions are not unique. An example is given below.

EXAMPLE 4.1
Consider the system

$$x(t+1) = \begin{pmatrix} -a & 1 \\ 0 & 1 \end{pmatrix} x(t) + \begin{pmatrix} b_1 \\ b_2 \end{pmatrix} u(t)$$

with the loss function

$$V = \sum_1^N x_1^2(t).$$

It is easy to show that if $|b_2| > |b_1|$ then the equation (4.2) has the following two positive solutions as $N \to \infty$.

$$S_1 = \begin{pmatrix} 1 & 0 \\ 0 & 0 \end{pmatrix}, \quad S_2 = \begin{pmatrix} s_1 & s_2 \\ s_2 & s_3 \end{pmatrix}$$

$$s_1 = 1 + \frac{a^2(b_2^2 - b_1^2)}{(ab_1 - b_2)^2}$$

$$s_2 = -\frac{a(b_2^2 - b_1^2)}{(ab_1 - b_2)^2}$$

$$s_3 = \frac{b_2^2 - b_1^2}{(ab_1 - b_2)^2} .$$

The corresponding feedback gains are

$$L_1 = \frac{1}{b_1} \begin{pmatrix} -a & 1 \end{pmatrix}$$

and

$$L_2 = \frac{ab_2 - b_1}{b_2(ab_1 - b_2)} \begin{pmatrix} -a & 1 \end{pmatrix}.$$

□

5. COMPARISON WITH MINIMUM VARIANCE CONTROL

The problem discussed in Chapter 2 can be regarded as a special case of the linear quadratic control problem. To see this consider a system with one input and one output described by (2.1). Let the criterion be

$$V = \min E \frac{1}{N} \sum_{t=1}^{N} y^2(t).$$

Change the coordinate system in such a way that the matrix A is in companion form. Applying the Theorem 3.1 it is then found that the equation (2.1) can be written as

$$\hat{x}(t+1) = \begin{pmatrix} -a_1 & 1 & \cdots & 0 \\ -a_2 & 0 & \cdots & 0 \\ \vdots & & & \\ -a_{n-1} & 0 & \cdots & 1 \\ -a_n & 0 & \cdots & 0 \end{pmatrix} \hat{x}(t) + \begin{pmatrix} b_1 \\ b_2 \\ \vdots \\ b_{n-1} \\ b_n \end{pmatrix} u(t) + \begin{pmatrix} k_1 \\ k_2 \\ \vdots \\ k_{n-1} \\ k_n \end{pmatrix} \varepsilon(t)$$

$$y(t) = \hat{x}_1(t) + \varepsilon(t). \tag{5.1}$$

It is easy to show by direct comparison that the relation between the input u and the output y can be written as

$$y(t) + a_1 y(t-1) + \ldots + a_n y(t-n) = b_1 u(t-1) + \ldots + b_n u(t-n) + \varepsilon(t) +$$
$$+ c_1 \varepsilon(t-1) + \ldots + c_n \varepsilon(t-n), \qquad (5.2)$$

where

$$c_i = a_i + k_i, \qquad i = 1, 2, \ldots, n. \qquad (5.3)$$

Equation (5.2) is, however, a CARMA model and the equivalence is thus obvious.

6. APPLICATIONS

The linear quadratic gaussian theory is frequently referenced in engineering literature and sometimes also in economics. It is, however, difficult to find good straightforward applications of the theory. Apart from the cases where minimum variance criteria apply it is not easy to find examples where the quadratic criterion (2.3) is well motivated physically. One rare case is the steering of ships where the average increase in drag due to steering can be expressed as

$$\frac{\Delta R}{R} = \frac{k}{T} \int_0^T [\psi^2(t) + \lambda \delta^2(t)] \, dt,$$

where ψ is the heading deviation and δ the rudder angle.

Another difficulty is to obtain appropriate models for the process dynamics and the environment. In spite of this it is frequently attempted to use the LQG theory to solve control problems because the structure of the solution is very appealing intuitively.

7. REFERENCES

The linear quadratic gaussian theory is well covered in textbooks [1], [2], [3], and [4]. A special issue [5] of the IEEE Transactions is entirely devoted to the theory and its applications. Application of the theory to steering of ships is discussed in [6]. Examples of the application of the theory to economic problems are found in [3].

[1]   B D O Anderson and J B Moore: Linear Optimal Control. Prentice Hall, Englewood Cliffs, N J, 1971.

[2]   K J Åström: Introduction to Stochastic Control Theory. Academic Press, New York, 1970.

[3]   G C Chow: Analysis and Control of Dynamic Economic Systems. John Wiley & Sons, New York, 1975.

[4]   H Kwakernaak and R Sivan: Linear Optimal Control Systems. Wiley--Interscience, New York, 1972.

[5]   M Athans (editor): Special Issue on the Linear-Quadratic-Gaussian Estimation and Control Problem. IEEE Trans AC-16 (1971), no. 6.

[6]   K J Åström: Some Aspects on the Control of Large Tankers. Proc Colloques IRIA Analyse de Systèmes et ses Orientations Nouvelles. Versailles-Rocquencourt, December 1976, to appear in Springer Lecture Notes.

CHAPTER 4 - CONTROL OF MARKOV CHAINS

1. INTRODUCTION

The previous chapters have dealt with linear systems only. In this chapter a nonlinear problem will be discussed. To make the analysis simple a model where the process dynamics is approximated by a markov chain will be investigated. The analysis follows the pattern of the previous chapters. The mathematical models of the process and its environment are discussed in Section 2. The solution to the filtering problem is given in Section 3 and the optimal control problem is solved in Section 4.

2. MATHEMATICAL MODELS

Internal descriptions on state models will be investigated. If it is attempted to generalize the model given by equation (2.1) in Chapter 3 to the nonlinear case it will be necessary to describe the conditional probability distribution of $x(t+1)$ given $x(t)$ and of the measurement $y(t)$ given $x(t)$. Such a description will be simplified if the state space is simple. It will therefore be assumed that the state vector x and the measurements y can assume finitely many values only. The stochastic process $\{x(t), t = 0, 1, \ldots\}$ thus becomes a markov chain.

It is assumed that the initial probability distribution of the states is given by

$$p_i^0 = \Pr\{x(0) = i\}, \quad i = 1, 2, \ldots, n. \tag{2.1}$$

The dynamic development of the state is described by the transition probability

$$p_{ij}(u,t) = \Pr\{x(t+1) = j | x(t) = i\}, \quad i,j = 1, 2, \ldots, n. \tag{2.2}$$

The transition probabilities may depend on time t and the control u. The transition probabilities have the properties

$$p_{ij}(u,t) \geq 0, \quad \sum_{j=1}^{n} p_{ij}(u,t) = 1. \tag{2.3}$$

The measurement process $\{y(t), t = 0, 1, \ldots\}$ is similarly characterized by the probabilities

$$q_{ij}(u,t) = \Pr\{y(t) = j | x(t) = i\}, \tag{2.4}$$

where

$$q_{ij}(u,t) \geq 0, \quad \sum_{j=1}^{m} q_{ij}(u,t) = 1. \tag{2.5}$$

The dynamics of the process and its environment are thus characterized by the matrices $P = \{p_{ij}, i,j = 1,\ldots,n\}$ and $Q = \{q_{ij}, i = 1,\ldots,n, j = 1,\ldots,m\}$ and by the initial distribution of the states.

It is assumed that the purpose is to control the system in such a way that the following loss function is as small as possible.

$$J = E \sum_{t=0}^{N} g[x(t), u(t), t],$$

where g is a function which assumes real values.

The admissible controls are assumed to be such that $u(t)$ is a function of $Y_t = [y(t), y(t-1), \ldots, y(0)]$ for each t.

3. OPTIMAL FILTERING

To solve the prediction problem we require the conditional probability distribution

$$w_i(t) = \Pr\{x(t) = i | Y_t\}. \tag{3.1}$$

If this probability distribution is known then many different predictors like the conditional mean, the value with highest probability etc. can easily be determined. In analogy with the linear case a recursive equation will be given for the predictor. This recursion is given by the following result.

THEOREM 3.1
Introduce the linear maps A_j defined by

$$(A_j w)_i = \sum_{k=1}^{n} q_{ij} p_{ki} w_k, \quad j = 1,\ldots,m \tag{3.2}$$

and introduce the norm

$$\| A_j w \| = \sum_{i=1}^{n} (A_j w)_i. \tag{3.3}$$

Then the conditional distribution $w(t)$ defined by (3.1) satisfies

$$w(t+1) = \frac{A_{y(t+1)} w(t)}{\| A_{y(t+1)} w(t) \|} \qquad (3.4)$$

and

$$\| A_j w(t) \| = \Pr\{y(t+1) = j | Y_t\}. \qquad (3.5)$$

Proof:
It follows from the multiplication rule for probabilities
$$\Pr\{x(t+1) = i | Y_{t+1}\} = \Pr\{x(t+1) = i | Y_t, y(t+1)\} =$$

$$= \frac{\Pr\{x(t+1) = i, y(t+1) = j | Y_t\}}{\Pr\{y(t+1) = j | Y_t\}}.$$

Furthermore the equations (2.2) and (2.4) give

$$\Pr\{x(t+1) = i, y(t+1) = j | Y_t\} = \sum_{k=1}^{n} q_{ij} p_{ki} w_k(t) = (A_j w)_i$$

and

$$\Pr\{y(t+1) = j | Y_t\} = \sum_{i=1}^{n} \sum_{k=1}^{n} q_{ij} p_{ki} w_k(t) = \| A_j w(t) \|$$

and the proof is complete. □

4. OPTIMAL CONTROL

Having solved the filtering problem the optimal control problem will now be discussed. A functional equation which characterizes the optimal solution will first be derived. The properties of the functional equation will then be discussed.

The Bellman Equation

Assume that the control u can take values in a finite set $\mathcal{U}$ only. The minimum of the loss function will then always exist. Introduce the function $V: R^n \to R$ defined by

$$V_t(w(t)) = \min_{u(t), u(t+1), \ldots} E\left\{ \sum_{k=t}^{N} g(x(k), u(k), k) | Y_t \right\}. \qquad (4.1)$$

Then

$$V_t(w(t)) = \min_u \left\{ \sum_{i=1}^n g(i,u,t) w_i(t) + \min_{u(t+1),\ldots} E\left[\sum_{t+1}^N g(x(k),u(k),k) | Y_t \right] \right\} =$$

$$= \min_u \left\{ <g(u,t), w(t)> + E\left[V_{t+1}(w(t+1)) | Y_t \right] \right\}.$$

where $<g,w>$ denotes a scalar product and $g(u,t)$ is a vector with components $g(i,u,t)$. It follows from Theorem 3.1 that conditioned on Y_t, $w(t+1)$ can assume m different values

$$w(t+1) = \frac{A_j w(t)}{\|A_j w(t)\|} \qquad j = 1,\ldots,m$$

with probabilities $\|A_j w(t)\|$. Hence

$$V_t(w(t)) = \min_u \left[<g(u,t), w(t)> + \sum_{j=1}^m \|A_j w(t)\| \, V_{t+1}(A_j w(t) / \|A_j w(t)\|) \right]. \tag{4.2}$$

We now have

THEOREM 4.1
A necessary condition for the minimum is that the function $V_t(w(t))$ satisfies the Bellman equation (4.2).

Proof:
It has been shown that if u is a minimizing feedback then there is a function V which satisfies (4.2). Conversely let v be an admissible control $v: R^n \to U$. Introduce the function W defined by

$$W_t(w(t)) = E\left\{ \sum_{k=t}^N g(x(k), v(w(k)), k) | Y_t \right\}.$$

Then W satisfies the recursion

$$W_t(w(t)) = <g(v,t), w(t)> + \sum_{j=1}^m \|A_j w(t)\| \, W_{t+1}(A_j w(t) / \|A_j w(t)\|).$$

It is now straightforward to show that

$$W_t(w) \geq V_t(w).$$

This is obviously true for $t = N$ and it follows for $t < N$ by induction. □

The equation (2.4), which is called the *Bellman equation*, plays the role of the Hamilton-Jacobi equation in stochastic control theory. When solving the Bellman equation two functions are obtained:

V: $R^n \to R$

and

u: $R^n \to U$ = {the set of possible controls}.

Structure of Optimal Feedback

The function u is a map from the conditional probability distributions to the controls. The structure of the optimal feedback is thus as shown in Fig. 6. It is composed of a filter which computes the conditional probability density of the states given the past controls and the past measurements. The filter is described by the equation (3.4). The other part of the feedback is the function u which is obtained from the solution of the Bellman equation.

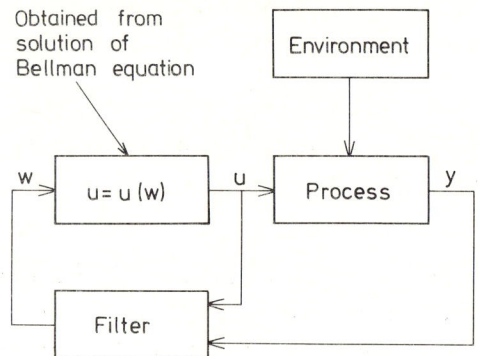

Fig. 6 - Block diagram of the optimal regulator. The filter computes the conditional distribution of the states given at past data y_t

Properties of the Bellman Equation

We have

THEOREM 4.2
The solution V of (4.2) is concave.

Proof:
An outline is given below. The full details are given in Åström [1]. The function

$$V_N = \min_u <g(u,t), w(t)>$$

is concave because it is the minimum of linear functions. Now use induction. Assume V_{t+1} concave. Then V_t is also concave because it is obtained by adding concave functions with positive weights and minimizing.
□

Computational Aspects

The Bellman equation can rarely be solved analytically. It is thus necessary to resort to numerical solutions. This is not trivial. To solve the equation (4.2) numerically it is necessary to store the functions V_t. This is a substantial burden if the number of states is large. Assume that there are n states. Because the components of w are probabilities the argument of the function can be characterized by n-1 variables in the range $0 \leq w_i \leq 1$. Assuming that the components of w are quantized in 10 levels each. It is then necessary to use 10^{n-1} cells to store the function V_t. For n = 11 the number is prohibitive even for the largest computers available today. It is then necessary to find good approximations of the function which are more economical storage wise.

The solution to the stochastic control problem is thus useful in the sense that it gives valuable insight into the structure of the optimal feedback. The solution is, however, not very practical in the sense that the computational effort to obtain the optimal feedback is prohibitive if the number of states is large.

5. AN EXAMPLE

An example is used to illustrate the properties of the solution. Consider a case where the transition and observation matrices are defined by

$$P = \begin{pmatrix} 0.9 & 0.1 \\ 0.1 & 0.9 \end{pmatrix}, \quad Q = \begin{pmatrix} 0.8 & 0.2 \\ 0.2 & 0.8 \end{pmatrix}, \quad u = 1$$

$$P = \begin{pmatrix} 0.8 & 0.2 \\ 0.2 & 0.8 \end{pmatrix}, \quad Q = \begin{pmatrix} 0.85 & 0.15 \\ 0.15 & 0.85 \end{pmatrix}, \quad u = 2$$

$$P = \begin{pmatrix} 0.6 & 0.4 \\ 0.4 & 0.6 \end{pmatrix}, \quad Q = \begin{pmatrix} 0.9 & 0.1 \\ 0.1 & 0.9 \end{pmatrix}, \quad u = 3$$

$$P = \begin{pmatrix} 0.4 & 0.6 \\ 0.6 & 0.4 \end{pmatrix}, \quad Q = \begin{pmatrix} 0.9 & 0.1 \\ 0.1 & 0.9 \end{pmatrix}, \quad u = 4$$

$$P = \begin{pmatrix} 0.2 & 0.8 \\ 0.8 & 0.2 \end{pmatrix}, \quad Q = \begin{pmatrix} 0.85 & 0.15 \\ 0.15 & 0.85 \end{pmatrix}, \quad u = 5$$

$$P = \begin{pmatrix} 0.1 & 0.9 \\ 0.9 & 0.1 \end{pmatrix}, \quad Q = \begin{pmatrix} 0.8 & 0.2 \\ 0.2 & 0.8 \end{pmatrix}, \quad u = 6.$$

The loss function is assumed to be given by

$$g = \begin{pmatrix} 1 & 0 \end{pmatrix}.$$

It is thus desired to keep the process in the second state. The conditional distribution can be chosen as the conditional probability for the process to be in state $\#$ 1. The Bellman equation was in this case solved numerically by quantizing this probability into 10 steps. The control law $u(w)$ obtained is given below.

step \ w	0.05	0.15	0.25	0.35	0.45	0.55	0.65	0.75	0.85	0.95
10	1	1	1	1	1	6	6	6	6	6
9	1	1	2	2	3	4	5	5	6	6
8	1	1	2	2	3	4	5	5	6	6
7	1	1	2	2	3	4	5	5	6	6
⋮										
1	1	1	2	2	3	4	5	5	6	6

The solution obtained for a finer quantization of 20 steps is given by

t \ w	0.025	0.075	0.125	0.175	0.225	0.275	0.325	0.375	0.425	0.475	0.525	0.575	0.625	0.675	0.725	0.775	0.825	0.875	0.925	0.975
10	1	1	1	1	1	1	1	1	1	1	6	6	6	6	6	6	6	6	6	6
9	1	1	1	1	2	2	2	2	3	3	4	4	5	5	5	5	6	6	6	6
8	1	1	1	2	2	2	2	3	3	3	4	4	4	5	5	5	5	6	6	6
7	1	1	1	2	2	2	2	2	3	3	4	4	5	5	5	5	5	6	6	6
6	1	1	1	2	2	2	2	3	3	3	4	4	4	5	5	5	5	6	6	6
5	1	1	1	2	2	2	2	2	3	3	4	4	5	5	5	5	5	6	6	6
4	1	1	1	2	2	2	2	2	3	3	4	4	5	5	5	5	5	6	6	6
3	1	1	1	2	2	2	2	2	3	3	4	4	5	5	5	5	5	6	6	6
2	1	1	1	2	2	2	2	2	3	3	4	4	5	5	5	5	5	6	6	6
1	1	1	1	2	2	2	2	2	3	3	4	4	5	5	5	5	5	6	6	6

In the last step t = 10 the solution is obvious if the probability of being in state # is less than 0.5, then the control #1 is chosen. This means that the probability of the state being unchanged is as large as possible. Otherwise control # 6 is chosen which means that the probability for a switch is maximized.

For step 9 the same policy is used provided that the probability of being in state # 1 is very small or very large. When the probability of being in state # is between 0.2 and 0.4 the optimal control is however u = 2. This means that the probability for a switch is higher than for u = 1. The measurements will, however, be more accurate which will benefit the conditional probabilities in the next step. If the probability of being in the state # 1 is between 0.4 and 0.5 it is beneficial to choose u = 3 which gives an even better measurement accuracy.

This example clearly illustrates some interesting properties of the solution to the nonlinear stochastic control problem. The control law may generate control actions that will drive the process away from its target provided that this will result in a more accurate estimate of the state. This property is called *dual control*.

Also notice in the tables above that the control law converges after a few steps only.

6. REFERENCES

The probabilistic model used in this section was introduced and analysed in [1]. The model is discussed extensively in [2]. The possibilities of approximating a system with a continuous state space by a markov chain is discussed in [3].

[1] K J Åström: Optimal Control of Markov Processes with Incomplete State Information, I and II. J Math Anal Appl 10 (1965) 174-205 and 26 (1969) 403-406.

[2] H J Kushner: Introduction to Stochastic Control. Holt, Rinehart and Winston, New York, 1971.

[3] R W Brockett: Stationary Covariance Generation with Finite State Markov Processes. Paper TA26-12:30, Joint Automatic Control Conference 1977, pp 1057-1060.

CHAPTER 5 - NONLINEAR STOCHASTIC CONTROL

1. INTRODUCTION

In this chapter the results of Chapter 4 will be extended to systems where the statespace is continuous. It will be shown that the problem formulation includes many interesting control problems. For example it is possible to treat adaptive control systems using these models. The theoretical results that are available are unfortunately fairly weak. Very little is known about existence. The order of presentation is the same as has been used in the previous chapters. The mathematical models used are discussed in Section 2. It is shown that the interesting cases of linear systems with drifting parameters and linear systems with constant but unknown parameters are included as special cases. The filtering problem is analysed in Section 3. A recursive equation is derived for the conditional probability density of the state variables. It is shown that the conditional densities are gaussian in particular cases. In Section 4 the control problem is investigated. The Bellman equation is derived formally. Unfortunately neither the recursive equation for the conditional density nor the Bellman equation are suitable to solve practical problems because of the excessive computational requirements. The analysis gives, however, interesting insight into the nature of the optimal solution. This insight can then be exploited to obtain different useful approximations. Some approximations are discussed in Section 5.

2. MATHEMATICAL MODELS

When analysing nonlinear problems it is frequently easier to work with internal descriptions. It is assumed as in Chapter 3 that the state x, the input u, and the output y take values in R^n, R^p, and R^r respectively. A general nonlinear generalization of the linear model discussed in Section 2 of Chapter 3 is then given by

$$\begin{cases} x(t+1) = f(x(t), u(t), v(t)) \\ y(t) = g(x(t), u(t), e(t)), \end{cases} \quad (2.1)$$

where $\{v(t)\}$ and $\{e(t)\}$ are sequences of random variables. The probability distribution of the initial state is characterized by

$$p^0(x) \, dx = \Pr \{x(t_0) \in x + dx\}, \quad (2.2)$$

where dx is an infinitesimal neighborhood of x.

To complete the characterization of the model it is necessary to specify the probability distributions of the disturbances v and e. If the model (2.1) should represent a state model in the sense that the conditional distributions of x(t+1) and y(t) given x(t) are the same as the conditional distributions of x(t+1) and y(t) given x(t), x(t-1), ... then {v(t)} and {e(t)} must be sequences of independent random variables. The stochastic process {x(t)} is then a Markov process. Instead of using the description (2.1) it is then natural to work directly with the probabilities

$$\begin{cases} p(\xi,x)\,dx = \Pr\{x(t+1) \in x + dx | x(t) = \xi\} \\ q(\xi,y)\,dy = \Pr\{y(t) \in y + dy | x(t) = \xi\}. \end{cases} \quad (2.3)$$

It is assumed that these densities exist. The densities p and q will also depend on t and u(t). This dependence is suppressed to simplify the notations. If the probability distributions for v(t) and e(t) are known the densities p and q can be determined. In the sequel it is therefore assumed that p and q are known.

The model (2.1) or (2.3) includes several special cases that are of great interest. Some will be discussed below.

EXAMPLE 2.1 (Linear Systems with Stochastic Parameters)
Consider a linear system characterized by the input-output relation

$$y(t+1) + a_1(t)y(t) + \ldots + a_n(t)y(t-n+1) = b_1(t)u(t) + \ldots + b_n(t)u(t-n+1) + e(t). \quad (2.4)$$

Introduce

$$\theta_1(t) = a_1(t), \ldots, \theta_n(t) = a_n(t), \theta_{n+1}(t) = b_1(t), \ldots, \theta_{2n}(t) = b_n(t)$$

and assume that the parameters are governed by

$$\theta(t+1) = \Phi\theta(t) + v(t), \quad (2.5)$$

where $\{v(t)\}$ is a sequence of uncorrelated gaussian random variables and the initial state $\theta(t_0)$ is gaussian (m, R_0). Introducing the vector

$$\varphi(t) = [-y(t) \ldots -y(t-n+1) \; u(t) \; u(t-1) \ldots u(t-n+1)]$$

the model (2.4) can then be written as

$$y(t+1) = \varphi(t)\theta(t) + e(t). \quad (2.4')$$

The system described by (2.4) and (2.5) is clearly of the form (2.1). A special case is when $\Phi = I$ which means that the parameters $a_i(t)$ and $b_i(t)$ are discrete Wiener processes. In specific applications it may not be realistic to assume that a model like (2.4) can be given for the parameter fluctuations. The case $\Phi = I$ can however serve as a generic case.

□

EXAMPLE 2.2 (Systems with Constant but Unknown Parameters)
It is frequently possible to assume that the equation which describes the process dynamics can be described by models having parameters which are constant but unknown. Such systems can be included into the form (2.1) by introducing the unknown parameters θ as extra state variables governed by the state equation

$$\theta(t+1) = \theta(t).$$

Consider for example a linear system governed by (2.4) where the parameters are constant. Such a system can be described by equations (2.4) and (2.5) with $\Phi = I$. The initial distribution of the state of (2.5) reflects the prior knowledge of the parameters.

□

3. OPTIMAL FILTERING

The optimal filtering problem will now be solved for the model (2.3). Since it is not possible to find a universally acceptable criterion the full conditional probability distribution of the state will be determined. It is thus assumed that the outputs $y(t), y(t-1), \ldots, y(t_0)$ have been observed. The problem is to determine the conditional probability distribution

$$\Pr\{x(t) \in A | Y_t\}, \qquad (3.1)$$

where Y_t is the values of all observed outputs or more precisely the σ-algebra generated by $y(t), y(t-1), \ldots, y(t_0)$ and $x(t_0)$. Assuming that the distribution (3.1) has a density denoted as

$$w(x,t)dx = \Pr\{x(t) \in x + dx | Y_t\},$$

we find from the properties of conditional densities that

$$w(x,t+1) = \frac{q(x,y(t+1)) \int p(\xi,x) w(\xi,t) d\xi}{\int\int q(x,y(t+1)) p(\xi,x) w(\xi,t) dx d\xi} \qquad (3.2)$$

$$w(x,t_0) = p^0(x).$$

The expression can be simplified somewhat if the linear positive operator A is defined by

$$(A_\eta w)(x) = q(x,\eta) \int p(\xi,x) w(\xi,t) d\xi . \tag{3.3}$$

Notice that A depends on u and t. Define the norm of a positive function w as

$$\|w(t)\| = \int w(x,t) dx . \tag{3.4}$$

Then the formula for updating the conditional density can be written as

$$w(\cdot,t+1) = A_{y(t+1)} w(\cdot,t) / \| A_{y(t+1)} w(\cdot,t) \| . \tag{3.5}$$

This formula is clearly a generalization of equation (3.4) in Chapter 4 for the markov chain case and we get

THEOREM 3.1
Consider a stochastic process defined by (2.2) and (2.3). The conditional density of the state $x(t)$ given past data Y_t is given by the recursive equation (3.5) with initial condition

$$w(t_0) = p^0 .$$

Furthermore

$$\Pr\{y(t+1) \in \eta + d\eta | Y_t\} = \| A_\eta w(\cdot,t) \| d\eta . \tag{3.6}$$

Even if the equation (3.5) looks innocent it requires extensive computations. Since w is a probability density over R^n the problem of storing the function is substantial. It is therefore interesting to consider special cases where the conditional density is simple. Such cases are discussed below.

THEOREM 3.2
Consider a linear system with random parameters described by (2.4) and (2.5) where $\{v(t)\}$ and $\{e(t)\}$ are sequences of independent gaussian $(0,R_1)$ and $(0,R_2)$ random variables. Let the initial parameter distribution be gaussian (m,R_0). Then the conditional distribution of the parameters $\theta(t)$ given Y_t is gaussian $(\hat{\theta}(t), P(t))$ where

$$\hat{\theta}(t+1) = \Phi \hat{\theta}(t) + K(t) [y(t+1) - \varphi(t) \hat{\theta}(t)]$$
$$K(t) = \Phi P(t) \varphi^T(t) [\varphi(t) P(t) \varphi^T(t) + R_2]^{-1}$$
$$P(t+1) = [\Phi - K(t) \varphi(t)] P(t) \Phi^T + R_1$$
$$\hat{\theta}(t_0) = m$$
$$P(t_0) = R_0 . \tag{3.7}$$

Furthermore the conditional density of $y(t+1)$ given Y_t is gaussian $(\varphi(t)\hat{\theta}(t), \sigma^2(t))$ where

$$\sigma^2(t) = R_2 + \varphi(t)P(t)\varphi^T(t). \tag{3.8}$$

Proof:
The system given by (2.4) and (2.5) is clearly a special case of (2.3). It follows from (2.4) that the density $p(\xi,x)$ is gaussian $(\Phi\xi, R_1)$ and that the density $q(\xi,y)$ is gaussian $(\varphi\xi, R_2)$. It was furthermore assumed that the prior density was gaussian (m, R_0). Repeated use of (3.2) then shows that the conditional density is also gaussian. The formulas (3.7) and (3.8) are verified simply by computing the density in the same way as was done when deriving the Kalman filter theorem in Section 3 of Chapter 3.
□

Remark 1
Notice that the distribution of $y(t)$ is not gaussian.

Remark 2
A similar result can be obtained for vector difference equations (2.3). In particular this means that if the parameters appear as elements of the matrices A and B in the model

$$x(t+1) = Ax(t) + Bu(t) + v(t),$$

then the conditional density of the parameters given $x(t), x(t-1), \ldots$ is also gaussian.

4. OPTIMAL CONTROL

The optimal control of processes described by equations (2.2) and (2.3) will now be investigated. It is assumed that the purpose of the control can be expressed as to minimize the loss function

$$J = E \sum_{t=t_0}^{N} h(x(t), u(t), t), \tag{4.1}$$

where $h: R^{n+p+1} \to R$. The admissible controls are assumed to be such that $u(t)$ is a function of $y(t), y(t-1), \ldots$ or more precisely that $u(t)$ is measurable with respect to the σ-algebra Y_t generated by $y(t), y(t-1), \ldots, y(t_0)$ and $x(t_0)$.

Proceeding in the same way as in the previous chapter it is first found that it is difficult to show existence of the minimum. It is therefore assumed that the minimum exists and we will proceed formally.

Introduce

$$V_t(w(\cdot,t)) = \min E\left\{\sum_{k=t}^{n} h(x(k), u(k), k) | Y_t\right\}. \tag{4.2}$$

We find

$$V_t(w(\cdot,t)) = \min_u \left\{\int h(\xi, u, t) w(\xi, t) d\xi + E\left[V_{t+1}(w(\cdot,t+1)) | Y_t\right]\right\}.$$

It follows from Theorem 3.1 that

$$V_t(w) = \min_u \left[<h, w> + \int \|A_\eta w\| V_{t+1}(A_\eta w / \|A_\eta w\|) d\eta\right], \tag{4.3}$$

where

$$<h, w> = \int h(\xi, u, t) w(\xi, t) d\xi.$$

The minimum is thus characterized by the Bellman equation (4.3). Notice that the argument of the Bellman equation is a density of a probability distribution over R^n. Even a numerical solution is thus not possible in the general case.

The convexity of V can be established in the same way as was done for systems with finite states in Chapter 4.

Even if the Bellman equation can not be solved the analysis shows that even in the general nonlinear case the structure of the feedback is that shown in Fig. 6 of Chapter 4. The optimal feedback can thus be thought of as being composed of two parts. One part is a nonlinear filter which computes the conditional density of the state vector given all observations. The other part is a nonlinear function which maps the conditional density on the control variables. This function can be precomputed from the Bellman equation.

5. LINEAR SYSTEMS WITH RANDOM PARAMETERS

In the general case the Bellman equation can neither be solved analytically nor computationally. Some special cases will therefore be investigated. The linear system with stochastic parameters given in Example 2.1 will be investigated.

It is assumed that the criterion is to minimize the loss function

$$J_N = E\left\{\sum_{t=t_0}^{N} [y(t) - y_r(t)]^2\right\}, \tag{5.1}$$

where $y_r(t)$ is a given reference value. The admissible controls are

assumed to be such that u(t) is a function of $Y_t = y(t),\ldots, y(t_0)$.
The reason for choosing the criterion (5.1) is that the solution to
the control problem is known if the parameters θ of the model are
known. In the case of constant parameter systems the solution was e.g.
given in Chapter 2. The analysis will thus illustrate the effects of
fluctuations in the model parameters.

Even in this simple case it is not easy to establish existence of
the minimum.

The Filtering Problem

The filtering problem will first be solved. The system is described by
equations (2.4) and (2.5). It can be brought to standard form (2.3)
by introducing a state composed of the vectors θ(t) and the vector
$\tilde{\varphi}$ which is defined by

$$\tilde{\varphi}(t) = [-y(t) \ldots -y(t-n+1) \quad 0 \quad u(t-1) \ldots u(t-n+1)]. \tag{5.2}$$

The conditional distribution of $\tilde{\varphi}(t)$ given Y_t is a point distribution. It was shown in Theorem 3.2 that the conditional distribution of
θ(t) given Y_t is gaussian $(\hat{\theta}(t), P(t))$ for linear systems with
random parameters. The conditional distribution of the state of the
system is called the *hyperstate*. It can be characterized by the triple

$$\xi(t) = [\tilde{\varphi}(t), \hat{\theta}(t), P(t)]. \tag{5.3}$$

The equation for updating $\tilde{\varphi}$ follows directly from (5.2). The
equations for updating $\hat{\theta}$ and P are given by Theorem 3.2.

The Control Problem

The filtering problem is thus easily solved for the particular model
structure chosen. To discuss the control problem we will first solve
the problem in the case the parameters are known. The problem will then
be solved for the special case when the criterion (5.1) only contains
one term. The solution of the complete problem is finally discussed.

Systems with Known Parameters

If the parameters of the system (2.4) are known it is easy to obtain
the optimal feedback. Introducing the vector $\tilde{\varphi}$ defined by (5.2) the
equation (2.4) can be written as

$$y(t+1) = \varphi(t)\theta(t) + e(t+1) = b(t)u(t) + \tilde{\varphi}(t)\theta(t) + e(t+1), \tag{5.4}$$

where we have introduced $b(t) = b_1(t)$. The optimal feedback is then given by

$$u(t) = \frac{y_r(t+1) - \tilde{\varphi}(t)\theta(t)}{b(t)}. \qquad (5.5)$$

The minimal loss is given by

$$\min J_N = (N+1-t_0)R_2.$$

Notice that it is necessary to impose the condition $b(t) \neq 0$ otherwise the control law (5.5) does not make sense. It is also necessary to require that the parameters $b_i(t)$ are such that the difference equation

$$b_1(t)u(t-1) + b_2(t)u(t-2) + \ldots + b_n(t)u(t-n) = 0$$

are asymptotically stable because otherwise the control signal will become infinitely large. Compare with the discussion in Section 4 of Chapter 2. A simple example is used as an illustration.

EXAMPLE 5.1
Consider a process described by

$$y(t+1) = y(t) + b(t)u(t) + e(t+1). \qquad (5.6)$$

This is a discrete time version of a continuous time system whose output is the time integral of its input. Let the reference value y_r be zero. If the parameter b is known then the control which minimizes J_1 or J_N is given by

$$u(t) = -\frac{y(t)}{b(t)}. \qquad (5.7)$$

The optimal feedback is thus a proportional regulator with gain $1/b(t)$.
□

Certainty Equivalence Control

When the parameters θ of the system (2.4) are not known it is tempting to replace the control law (5.5) by the control law

$$u(t) = \frac{y_r(t+1) - \tilde{\varphi}(t)\hat{\theta}(t)}{\hat{b}(t)}, \qquad (5.8)$$

where $\hat{\theta}(t)$ is the conditional mean of the values $\theta(t)$ of the unknown parameters given Y_t.

EXAMPLE 5.2
The certainty equivalence control for the process (5.6) is given by

$$u(t) = -\frac{1}{\hat{b}(t)}y(t). \qquad (5.9)$$

□

Cautious Control or One-Step Control

The special case when the criterion (5.1) has one term only is first discussed. According to Theorem 3.2 the conditional distribution of $y(t+1)$ given Y_t is gaussian with mean $\varphi(t)\hat{\theta}(t)$ and covariance $R_2 + \varphi(t)P(t)\varphi^T(t)$. Then

$$E\{[y_r(t+1) - y(t+1)]^2 | Y_t\} = [y_r(t+1) - \varphi(t)\hat{\theta}(t)]^2 + R_2 + \varphi(t)P(t)\varphi^T(t). \quad (5.10)$$

To see how the right hand side depends on the control $u(t)$, introduce equation (5.4). Then

$$E\{[y_r(t+1) - y(t+1)]^2 | Y_t\} = R_2 + [y_r(t+1) - \hat{b}(t)u(t) - \tilde{\varphi}(t)\hat{\theta}(t)]^2 +$$
$$+ u^2(t)p_b(t) + 2u(t)\tilde{\varphi}(t)P(t)\ell + \tilde{\varphi}(t)P(t)\tilde{\varphi}^T(t), \quad (5.11)$$

where ℓ is a column-vector which selects the $(n+1)$:th row of the matrix P i.e.

$$\ell = \text{col}\,(\underbrace{0 \ldots 0}_{n}\; 1\; \underbrace{0 \ldots 0}_{n-1}).$$

Minimization of (5.11) with respect to $u(t)$ gives

$$\min_{u(t)} E\{[y_r(t+1) - y(t+1)]^2 | Y_t\} = R_2 + [y_r(t+1) - \tilde{\varphi}(t)\hat{\theta}(t)]^2 + \tilde{\varphi}(t)P(t)\tilde{\varphi}^T(t)$$
$$+ \frac{[y_r(t+1) - \hat{b}(t)\tilde{\varphi}(t)\hat{\theta}(t) - \tilde{\varphi}(t)P(t)\ell]^2}{\hat{b}^2(t) + p_b(t)}, \quad (5.12)$$

where the minimum is attained for the control law

$$u(t) = \frac{y_r(t+1) - \hat{b}(t)\tilde{\varphi}(t)\hat{\theta}(t) - \tilde{\varphi}(t)P(t)\ell}{\hat{b}^2(t) + p_b(t)}. \quad (5.13)$$

This control law is called *one-step control*, because it minimizes the expected loss over one step only. Notice that in the case of known parameters the minimal loss is a constant. This means that the one-step control is optimal also for the N-step criterion. This is, however, not the case when the parameters are unknown.

EXAMPLE 5.3
The one-step control for the process (5.6) is given by

$$u(t) = -\frac{\hat{b}(t)}{\hat{b}^2(t) + p_b(t)} y(t) = -\frac{\hat{b}^2(t)}{\hat{b}^2(t) + p_b(t)} \cdot \frac{1}{\hat{b}(t)} y(t). \quad (5.14)$$

□

For the control laws, ((5.5) known parameters), ((5.8) certainty equivalence), and ((5.13) one-step), the input-output relation can be expressed as

$$u(t) = \alpha_1(t)y(t) + \ldots + \alpha_n y(t-n+1) + \beta_1(t)u(t-1) + \ldots + \beta_n(t)u(t-n+1).$$

When the parameters are known (5.5) the coefficients $\alpha_i(t)$ and $\beta_i(t)$ are simply functions of time. But when the parameters are stochastic the parameters $\alpha_i(t)$ and $\beta_i(t)$ will depend also on past observations Y_t.

Notice that the one-step control law (5.13) reduces to the certainty equivalence control if the conditional variance $p_b(t)$ of the estimate $\hat{b}(t)$ is zero.

The examples 5.2 and 5.3 clearly illustrate the differences between the one-step and the certainty equivalence controls. In these examples both regulators are simply proportional controllers. The one-step control (5.13) has a gain which is a factor

$$\hat{b}^2(t) / [\hat{b}^2(t) + p_b(t)]$$

lower than the gain of the certainty equivalence control. The effect of the parameter uncertainty is thus to reduce the gain. For this reason the one-step control is also called the *cautious* regulator.

The cautious regulator does not have the dual property discussed in Section 5 of Chapter 4. To obtain such a regulator it is necessary to solve the multistep optimization problem.

Multistep Optimization

To solve the multistep optimization problem it is necessary to solve the Bellman equation for the stochastic control problem. The Bellman equation can be simplified because the conditional distributions are gaussian.

Assume that the minimum exists and recall that the conditional distribution of the state given the measurements can be characterized by the triplet (5.3). Introduce

$$V(\tilde{\varphi}(t), \hat{\theta}(t), P(t), t) = \min_u E\left\{\sum_{k=t+1}^{N} [y(k) - y_r(k)]^2 \mid Y_t\right\}.$$

The following recursive equation is then obtained.

$$V(\tilde{\varphi}(t), \hat{\theta}(t), P(t), t) = \min_{u(t)} E\left\{[y(t+1) - y_r(t+1)]^2 \right.$$
$$\left. + V(\tilde{\varphi}(t+1), \hat{\theta}(t+1), P(t+1), t+1) \mid Y_t\right\}. \quad (5.15)$$

To proceed it is necessary to have the equations for the updating of the hyperstate (5.3). It follows from (2.4') and (5.2) that

$$\tilde{\varphi}(t+1) = \tilde{\varphi}^T(t)C + f\,\varphi(t)\hat{\theta}(t) + f\,\sigma(t)\varepsilon(t+1),$$

where ε is the normalized innovation

$$\varepsilon(t+1) = [y(t+1) - \varphi(t)\hat{\theta}(t)] / \sigma(t),$$

$$\sigma^2(t) = \varphi(t)P(t)\varphi^T(t) + R_2,$$

and

$$C = \begin{pmatrix} S_1 & 0 \\ 0 & S_1 \end{pmatrix}, \quad S_1 = \begin{pmatrix} 0 & 1 & 0 & \cdots & 0 \\ 0 & 0 & 1 & \cdots & 0 \\ \vdots & & & & \\ 0 & 0 & 0 & \cdots & 1 \\ 0 & 0 & 0 & \cdots & 0 \end{pmatrix}, \quad f = \begin{pmatrix} 1 & 0 & \cdots & 0 \end{pmatrix}.$$

It also follows from (3.7) that

$$\hat{\theta}(t+1) = \Phi\hat{\theta}(t) + K(t)\sigma(t)\varepsilon(t+1).$$

The conditional densities of $\tilde{\varphi}(t+1)$ and $\hat{\theta}(t+1)$ given Y_t are thus gaussian. Furthermore the conditional distribution of $P(t+1)$ given Y_t is a point distribution with all mass in

$$P(t+1) = [\Phi - K(t)\varphi^T(t)]P(t)\Phi^T + R_1,$$

where

$$K(t) = \Phi P(t)\varphi^T(t)\sigma^{-2}(t).$$

The functional equation (5.15) can then be written as

$$V(\varphi, \hat{\theta}, P, t) = \min_u \left\{ [y_r(t+1) - \varphi\hat{\theta}]^2 + \sigma^2 + \frac{1}{\sqrt{2\pi}} \int_{-\infty}^{\infty} e^{-\varepsilon^2/2} \right.$$

$$\left. \cdot V[\tilde{\varphi}C + f\varphi\hat{\theta} + f\sigma\varepsilon,\ \Phi\hat{\theta} + K\sigma\varepsilon,\ (\Phi - K\varphi^T)P\Phi^T + R_1,\ t+1]d\varepsilon \right\}, \quad (5.16)$$

where

$$\varphi\hat{\theta} = \tilde{\varphi}\hat{\theta} + \hat{b}u$$

$$K = \Phi P\varphi^T\sigma^{-2}$$

$$\sigma^2 = \varphi P\varphi^T + R_2.$$

Notice that φ, K, P, and σ depend on the control u.

The equation (5.16) is the Bellman equation for the problem. The equation can not be solved analytically. The equation can be solved numerically if the order of the system is sufficiently small.

EXAMPLE 5.4
For the system described by (5.6) the conditional distribution can be characterized by three variables $y(t)$, $\hat{b}(t)$, and $P(t)$. Assuming that $\Phi = 1$ and $R_2 = r_2$ the Bellman equation (5.16) then becomes

$$V(y,\hat{b},P,t) = \min_u \left\{ [y_r(t+1) - y - \hat{b}u]^2 + r_2 + u^2 P + \frac{1}{\sqrt{2\pi}} \int_{-\infty}^{\infty} e^{-\varepsilon^2/2} \right.$$

$$\left. \cdot V\left(y + \hat{b}u + \sqrt{r_2+u^2 P}\, \varepsilon,\ \hat{b} + \frac{uP}{\sqrt{r_2+u^2 P}}\, \varepsilon,\ \frac{r_2 P}{r_2+u^2 P} + r_1,\ t+1\right) d\varepsilon \right\}.$$

(5.17)

□

The Bellman equation has been solved numerically for special low order examples. The solutions have given insight into the nature of the optimal strategies. In particular it has been found that the character of the multistep optimization is different from the cautious control in the sense that the optimal feedback is *dual*. The optimal feedback will thus generate control actions which will improve the accuracy of future estimates at the expense of increased short term loss. The properties of the optimal feedback are thus similar to those found for the markov chain example in Section 5 of Chapter 4. Another interesting property is that the optimal feedback may be discontinuous in $P(t)$. From a theoretical point of view there are, however, many important problems that still remain unsolved. The existence of stationary policies as $N \to \infty$, the existence of optimal solutions are typical examples.

Because of the difficulty of solving the Bellman equation several suboptimal control strategies have been proposed. They have mainly been investigated by simulations.

6. REFERENCES

The concept of dual control was introduced by Feldbaum [1]. The Bellman equation can be regarded as a natural extension of classical Hamilton-Jacobi theory to stochastic control problems. See [2]. This chapter is largely based on the paper [3] which also contains an example of a dual control law obtained by solving the Bellman equation. Further examples of this type are given in [4] and [5]. Different suboptimal controls are discussed in [6] which contains many additional references and several simulation examples.

[1] A A Feldbaum: Dual Control Theory I-IV. Automation and Remote Control $\underline{21}$ (1961) 874-880, 1033-1039, and $\underline{22}$ (1961) 1-12, 109-121.

[2] R Bellman: Adaptive Control Processes - A Guided Tour. Princeton Univ Press, Princeton, 1961.

[3] K J Åström and B Wittenmark: Problems of Identification and Control. J Math Anal Appl $\underline{34}$ (1971) 90-113.

[4] T Bohlin: Optimal Dual Control of a Simple Process with Unknown Gain. Technical Paper PT 18.196, IBM Nordic Laboratory, Lidingö, Sweden, 1969.

[5] O L R Jacobs and J W Patchell: Caution and Probing in Stochastic Control. Int J on Control $\underline{16}$ (1972) 189-199.

[6] J Sternby: Topics in Dual Control. PhD dissertation, Dept of Automatic Control, Lund Institute of Technology, CODEN: LUTFD2/(TFRT-1012)/1-135/(1977).

CHAPTER 6 - SELF-TUNING REGULATORS

1. INTRODUCTION

It was shown in Chapters 2 and 3 that optimal stochastic control problems could be formulated and solved at least for linear systems with quadratic criteria. It is, however, a difficulty from the point of view of applications that the models describing the process and its environment are rarely known apriori. Nonlinear stochastic control problems were formulated in Chapters 4 and 5. It was shown that linear systems with unknown parameters could fit into the problem formulation provided that the state was measured exactly or that the dynamics could be described by a controlled autoregression. The theory developed in Chapter 5 can thus be applied to generalize some linear problems to the case of unknown parameters. The results of Chapter 5 give interesting insight into the structure of the optimal feedback. The results are, however, discouraging from a practical point of view because of the computations required to obtain the optimal feedback. It is thus meaningful to attempt a reformulation of the control problem which will lead to practical solutions. The following is one possibility. Consider a system with constant but unknown parameters and a criterion. Find a control law, which only operates on past input-output data, which does not require knowledge of the system parameters, and which converges to the optimal regulator that could be designed if the parameters of the process were known. Such a regulator is called a *self-tuning* regulator. There are many self-tuning regulators. The optimal dual controller is clearly self-tuning. It will be shown in this chapter that there are indeed self-tuning regulators at least for simple problems. The mathematical model of the process and its environment is discussed in Section 2. To keep the analysis simple only a simple first order system is treated. The problem formulation is also given in Section 2. In Section 3 it is shown that a solution is given by a comparatively simple feedback law. The properties of this feedback are analysed in Section 4.

2. MATHEMATICAL MODEL

It is assumed that the dynamics of the process and its environment can be described by the simple first order system

$$y(t) + ay(t-1) = bu(t-1) + e(t) + ce(t-1), \qquad (2.1)$$

where u is the control variable, y the output and $\{e(t)\}$ a sequence of independent gaussian random variables. It is furthermore assumed that the criterion is to minimize the quadratic loss function

$$J = \lim_{N \to \infty} E \frac{1}{N} \sum_{t=1}^{N} y^2(t). \tag{2.2}$$

The admissible controls are assumed such that $u(t)$ is a function of all past outputs $y(t), y(t-1), \ldots$. If the parameters are known it follows from Theorem 4.1 of Chapter 2 that the optimal control is the proportional feedback

$$u(t) = \frac{a-c}{b} y(t). \tag{2.3}$$

The problem of finding a self-tuning regulator can be stated as follows. Find a feedback law which does not depend on knowledge of the parameters a, b, and c which converges to the control law (2.3) as time increases.

3. A SIMPLE SELF-TUNING REGULATOR

If it is attempted to solve the problem formulated in Section 2 using the methods discussed in Chapter 5 the filtering problem must first be solved. The state of the system is $y(t)$ and the parameters a, b, and c. For the filtering problem a prior distribution for a, b, and c must be assumed. Recursive equations for the conditional distribution of a, b, and c given y_t can then be obtained. This distribution can be simplified a little by observing that the conditional distribution of a and b given c and y_t is gaussian. It can be characterized by two mean values $(\hat{a}, \hat{b})$ and three covariances $(p_a, p_{ab}, \text{ and } p_b)$. The conditional distribution of a, b, and c given y_t can thus be characterized by the conditional distribution of c and 5 real variables $(\hat{a}, \hat{b}, p_a, p_{ab}, \text{ and } p_b)$. The problem is simplified considerably if the parameter c is known because the conditional density of a and b given y_t is then gaussian as was shown in Theorem 3.2 of Chapter 5.

Assuming that the filtering problem is solved it can then be attempted to solve the Bellman equation. This may perhaps be done computationally in the case c is known because the state can then be characterized by 5 real variables. It is thus clear that even in a simple case like this it is not possible to compute the optimal dual control law if c is unknown.

Since the parameters of the process are constant it can be expected that the conditional distributions of the parameters a, b, and c given Y_t will converge to point distributions. For large t it can then be expected that reasonably good self-tuning strategies can be obtained from control laws that are computed from parameter estimates only.

It will be shown that there are indeed simple self-tuning control laws. One possibility is given by the following control law

$$u(t) = \hat{\theta}(t) y(t), \qquad (3.1)$$

where $\hat{\theta}(t)$ is the least squares estimate of the parameter θ in the model

$$y(t) + \theta y(t-1) = u(t-1) + e(t) \qquad (3.2)$$

based on data available up to time t i.e. $y(t), y(t-1), \ldots, y(1)$, $u(t-1), u(t-2), \ldots, u(1)$. The least squares estimate $\hat{\theta}$ is given by

$$\hat{\theta}(t) = -\left[\sum_{k=1}^{t-1} [y(k+1) - u(k)] \, y(k) \right] \Big/ \sum_{k=1}^{t-1} y^2(k). \qquad (3.3)$$

The control algorithm given by (3.1) and (3.3) can be expected to work nicely for the system (2.1) if c = 0 and b = 1. In this case the least squares estimate $\hat{\theta}$ will converge to a as $t \to \infty$ and the control law (3.1) will converge to

$$u(t) = ay(t),$$

which is the desired control law.

It is a remarkable property of the feedback law described by (3.1) and (3.3) that it will converge to the optimal law (2.3) also when $c \neq 0$. This is illustrated in the following example.

EXAMPLE 3.1
Consider a system (2.1) with a = -1, c = -0.7, and b = 1. With these numerical values the optimal feedback is

$$u(t) = -0.3 \, y(t).$$

Fig. 7 shows a simulation of the regulator (3.1), (3.3) applied to the system. From the simulation it appears that the parameter estimate converges to the value $\hat{\theta} = -0.3$ (which is the gain of the optimal feedback).

To compare the self-tuning regulator (3.1) (3.3) with the optimal regulator in the case of known parameters the accumulated loss defined by

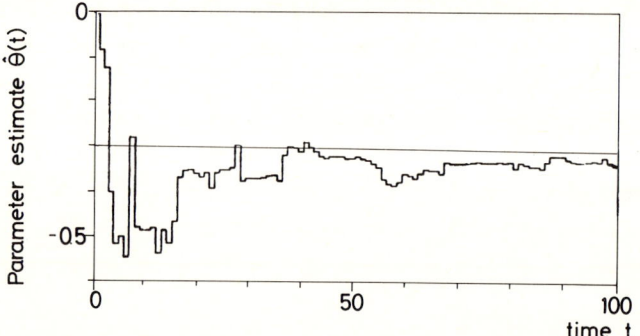

Fig. 7 - Parameter estimate $\hat{\theta}$ obtained in a simulation of the self-tuning regulator (3.1), (3.3) applied to the system (2.1) with a = -1, c = -0.7, and b = 1

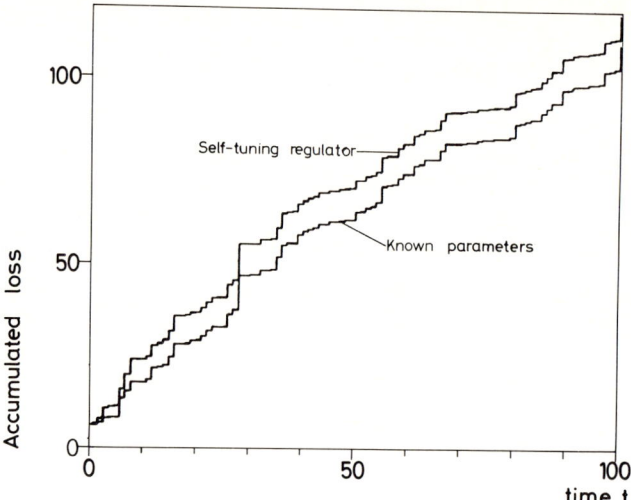

Fig. 8 - Accumulated loss functions for the self-tuning regulator (3.1), (3.3) and the optimal regulator based on known parameters

$$V(t) = \sum_{k=1}^{t} y^2(k)$$

has been calculated for the self-tuning regulator and the optimal regulator

u(t) = -0.3 y(t).

The results are shown in Fig. 8. It is clear from this figure that there is not a large difference between the performance of the two regulators. In particular it is seen in Fig. 8 that the loss functions will be virtually the same if the first 20 steps are neglected.

□

The example shows that the simple self-tuning feedback (3.1), (3.3) will perform very well. After a short transient it will give almost the same performance as a regulator based on exact knowledge of the system parameters. It is therefore of interest to analyse the performance of the simple regulator described by the equations (3.1) and (3.3).

4. ANALYSIS

The control law (3.1), (3.3) is a nonlinear feedback. It will now be analysed what happens when the regulator is applied to a process described by

$$y(t+1) + ay(t) = u(t) + n(t), \tag{4.1}$$

where n is a disturbance.

If the disturbance is bounded in the sense

$$\frac{1}{t} \sum_{k=1}^{t} n^2(k), \tag{4.2}$$

then the mean square output of the closed loop system

$$\frac{1}{t} \sum_{k=1}^{t} y^2(k) \tag{4.3}$$

is also bounded.

This statement is shown by contradiction. Assume therefore that (4.3) is not bounded. Equations (3.3) and (4.1) give

$$\hat{\theta}(t+1) = -\left[\sum_{k=1}^{t} [y(k+1) - u(k)] \, y(k)\right] \Big/ \sum_{k=1}^{t} y^2(k) = a - \sum_{k=1}^{t} n(k)y(k) \Big/ \sum_{k=1}^{t} y^2(k).$$

Schwartz' inequality gives

$$|\hat{\theta}(t+1) - a| \leq \sqrt{\left[\frac{1}{t}\Sigma n^2(k)\right]\left[\frac{1}{t}\Sigma y^2(k)\right]} \Big/ \frac{1}{t}\Sigma y^2(k) = \sqrt{\frac{1}{t}\Sigma n^2(k)} \Big/ \sqrt{\frac{1}{t}\Sigma y^2(k)}.$$

Since (4.2) is bounded and since it was assumed that (4.3) was unbounded then $\hat{\theta}$ is arbitrarily close to a for large t. The closed loop system is then given by

$$y(t+1) + [a - \hat{\theta}(t)] \, y(t) = n(t).$$

The solution to this linear equation can be written as

$$y(t) = \psi(t, t_0) y(t_0) + \sum_{k=t_0}^{t-1} \psi(t, k+1) n(k),$$

where the fundamental solution ψ is given by

$$\psi(t, k) = \begin{cases} 1 & k = t \\ \prod_{i=k}^{t-1} [a - \hat{\theta}(i)] & k < t. \end{cases}$$

Since $\hat{\theta}(t)$ is arbitrarily close to a for large t, we have

$$|\psi(t,k)| < \varepsilon^{t-k}, \quad t > k.$$

If n is bounded in the mean square sense then y is also bounded in the mean square and we have a contradiction.

The self-tuning regulator (3.1), (3.3) will thus always stabilize the system (4.1) in the mean square sense. It can be shown in this case that the parameter estimate $\hat{\theta}(t)$ will also converge. If $\hat{\theta}(t)$ converges as $t \to \infty$ it is easy to find the convergence point. The normal equations can be written

$$\sum_{k=1}^{t} y(k+1) y(k) = \hat{\theta}(t+1) \sum_{k=1}^{t} y^2(k) + \sum_{k=1}^{t} y(k) u(k).$$

Equation (3.1) now gives

$$\frac{1}{t} \sum_{k=1}^{t} y(k+1) y(k) = \frac{1}{t} \sum_{k=1}^{t} [\hat{\theta}(t+1) - \hat{\theta}(k)] y^2(k).$$

The right member converges to zero as $t \to \infty$ because $\hat{\theta}(t)$ converges and (4.3) is bounded. It is thus shown that if the parameter estimates converge then

$$\lim_{t \to \infty} \frac{1}{t} \sum_{k=1}^{t} y(k+1) y(k) = 0. \tag{4.4}$$

The self-tuning regulator (3.1), (3.3) thus attempts to make the correlation of the closed system output zero at lag 1. Assuming that the process to be controlled is given by (2.1) it now follows that it is only one value of $\hat{\theta}$ for which (4.3) is bounded and (4.4) holds, namely

$$\theta = a - c.$$

It has thus been established that the regulator (3.1), (3.3) is self--tuning for the system (2.1) and the minimum variance criterion. The analysis can be extended to the case when $b \neq 1$. The condition required is that $0 < b < 2$. The results can be extended to control of an n:th order CARMA process. Additional conditions are then required both for stability and convergence. There are also cases where the parameter estimates do not converge.

5. CONCLUSIONS

The self-tuning regulator given by (3.1) and (3.3) is much simpler than the optimal dual regulator. The performance of the dual regulator will be better than the self-tuning regulator. It will, however, be worse than the performance of the regulator based on exact knowledge of the system parameters. In the simple example there is a difference in the transient performances of the different regulators say in the first 20 steps in Fig. 8. After about 20 steps there is, however, little difference between the accumulated loss of the self-tuning regulator and the regulator based on exact knowledge of the parameters and consequently little room for improvement.

There are many different possibilities to design self-tuning regulators. Other recursive parameter estimation schemes than least squares can be used. They can be combined with many different procedures for control design. It is also possible to take uncertainties of the estimates into account and also to incorporate some approximative dual control features. Some progress has been made towards the analysis and understanding of such control laws. There are, however, still many open problems.

6. REFERENCES

Self-tuning regulators have been used to control industrial processes. Several applications are given in the paper [1] which also contains a review of the theory and many references. The convergence analysis is given in [2].

[1] K J Åström, U Borisson, L Ljung, and B Wittenmark: Theory and Applications of Self-Tuning Regulators. Automatica 13 (1977) 457-476.

[2] L Ljung: Analysis of Recursive Stochastic Algorithms. IEEE Trans AC-22 (1977), 551-575.

LIE THEORY, FUNCTIONAL EXPANSIONS, AND NECESSARY CONDITIONS IN SINGULAR OPTIMAL CONTROL

Roger W. Brockett

1. Introduction

In recent years there has been considerable interest in exploring the kind of results which can be obtained using some basic ideas from differential geometry, Lie groups, and so on, in the study of controlled dynamical systems. By now quite a few separate areas have been investigated (see, for example, [1], [2], [6], [8], [9], [10], [12], [15], [16], [17], [18], [19]) and a summary would be unwieldy. Instead, our purpose in this set of notes is to develop some ideas which center around singular optimal control and which may be treated by differential geometric methods. In doing so we will have occasion to use ideas from nonlinear system theory in a substantial way. Our main tool will be a function space Taylor series expansion which has recently been developed and which, in essence, allows one to eliminate the differential equation constraint from control problems. The properties of this expansion are, in turn, intimately connected with properties of the Lie algebras of vector fields generated by the system.

What we do here is to provide a means of classifying singular control problems in accordance with the "depth" of the necessary conditions which are available. In this theory one needs a certain amount of smoothness. The exact amount depends on the degeneracy of the problem. In fact, as a corollary of this point of view we can give a suitable definition for what one should mean by a "smooth problem" in optimal control theory.

2. Some Notation

We will have considerable use for power series expansions and in this connection we want to use a notation which permits an easy manipulation of multivariable Taylor series. If $x = \text{col}(x_1, x_2, \ldots, x_n)$ then we introduce $x^{[p]}$ as the $\binom{n+p-1}{p}$-tuple of independent monomials of degree p in $\{x_1, x_2, \ldots, x_n\}$, normalized in such a way as to make certain estimates work out in a natural way. Specifically, we choose $\alpha_{ij\ldots k} > 0$ such that for

$$x^{[p]} = \begin{bmatrix} \alpha_{11\ldots 1} x_1^p \\ \alpha_{21\ldots 1} x_2 x_1^{p-1} \\ \ldots \\ \alpha_{nn\ldots n} x_n^p \end{bmatrix}$$

we have $\langle x, x \rangle^{2p} = \langle x^{[p]}, x^{[p]} \rangle$ where $\langle \, , \, \rangle$ indicates ordinary inner product.

If $f : \mathbb{R}^n \to \mathbb{R}^m$ has r continuous derivatives in a neighborhood of x_0 then we can write

$$f(x) = f(x_0) + F_1(x-x_0) + F_2(x-x_0)^{[2]} + \ldots + F_r(x-x_0)^{[r]} + \phi(x)$$

where the L_i are matrices of dimension m by $\binom{n+i-1}{i}$ and $\phi(x)/\|x-x_0\|^r$ approaches zero with $x - x_0$.

If $u : \mathbb{R}^1 \to \mathbb{R}^m$ then we will be interested in expansions of the form

$$F(u) = F(0) + \int_0^t W^1_{\sigma_1}[u(\sigma_1)] d\sigma_1 + \int_0^t \int_0^{\sigma_1} W^2_{\sigma_1 \sigma_2}[u(\sigma_1) \otimes u(\sigma_2)] d\sigma_1 d\sigma_2 + \ldots$$

where W^i are linear functions of $u(\sigma_1) \otimes u(\sigma_2) \otimes \ldots \otimes u(\sigma_i)$. To simplify notation we write $u^{[i]}(\sigma_1, \sigma_2, \ldots, \sigma_i)$ to indicate this tensor product. Since the W's appear under integrals we can symmetrize or triangularize to make them unique. Our preference is to ask that

$$W^i_{\sigma_1 \sigma_2 \ldots \sigma_i} = 0 \quad \text{unless} \quad \sigma_1 \geq \sigma_2 \geq \ldots \geq \sigma_i .$$

3. Volterra Expansions

We deal here with systems which in local coordinates take the form

$$\dot{x}(t) = f[x(t), t] + \sum_{i=1}^{m} u_i(t) g_i[x(t), t] \; ; \; y(t) = h[x(t), t] \; ; \; x(0) = x_0 \; . \quad (*)$$

Globally we have a paracompact Hausdorff manifold M and vector fields F and $\{G_i\}_{i=1}^{m}$ defined on M. We assume that F is complete and that $h : M \to R$ is smooth. For each piecewise continuous choice of $u(\cdot) = (u_1(\cdot), u_2(\cdot), \ldots, u_m(\cdot))$ on $[0, T]$ we have an integral curve of $(*)$ defined in a neighborhood of $t = 0$. It may not exist on $[0, T]$ because only f is assumed to be complete. The first point we wish to discuss here is the existence of a convergent expansion for y in terms of $u(\cdot)$ taking the form

$$y(t) = y_0(t) + \int_0^t W_1(t, \sigma) u(\sigma) d\sigma + \int_0^t \int_0^{\sigma_1} W_2(t, \sigma_1, \sigma_2) u^{[2]}(\sigma_1, \sigma_2) d\sigma_1 d\sigma_2 + \ldots \; .$$

Let $C^m[0, t]$ denote the space of continuous $\mathbb{R}^m$ valued functions on $[0, t]$ normed with the sup norm. For each fixed value of t this would be the Taylor series expansion of the function space mapping from $C^m[0, t] \times C^m[0, t] \times \ldots \times C^m[0, t]$ into R. If we think of $(*)$ as defining a variational problem – minimize $y(t)$ subject to $(*)$ – then certain necessary conditions for $u(t) \equiv 0$ to be minimizing are obvious from the expansion.

In [4] we proved that if f, g and h are analytic then for each fixed T there exists $\varepsilon(T)$ such that $(*)$ possesses a Volterra series expansion which converges on $[0, T]$ for all piecewise continuous $u(\cdot)$ with norm less than $\varepsilon(T)$; only the case where $\dim u = 1$ was treated. In [5] we indicated the modification needed to cover the case of vector inputs and, more importantly, indicated how one could establish estimates of the form

$$y(t) = y_0(t) + \int_0^t W_{\sigma_1}^1 u(\sigma_1) d\sigma_1 + \ldots +$$

$$\int_0^t \int_0^{\sigma_1} \ldots \int_0^{\sigma_{p-1}} W_{\sigma_1 \sigma_2 \ldots \sigma_p}^p u^{[p]} d\sigma_1 d\sigma_2 \ldots d\sigma_p + \phi(t)$$

where for

$$\sup_{0 \leq \sigma \leq t} |u(\sigma)| \leq \varepsilon$$

we have

$$\lim_{\varepsilon \to 0} \frac{\phi(t)}{\varepsilon^p} = 0 \; ,$$

provided that f and h have p continuous derivatives and g has $p - 1$ continuous derivatives.

In [4] there is an algorithm for computing the kernels w^i which displays the dependence of the w^i on the Taylor series expansions of f, g, and h. This algorithm proceeds in two steps. The first is to note that in the special case

$$\dot{x}(t) = A(t)x(t) + \sum_{i=1}^{m} u_i(t)B_i(t)x(t)$$

one can make a change of variables such that $z(t) = P(t)x(t)$ and

$$\dot{z}(t) = \sum_{i=1}^{m} u_i(t)\tilde{B}_i(t)z(t) .$$

In this coordinate system the Peano Baker series is actually a Volterra series in the u's. The second step is to construct an approximation of order k to (*) which is of this "bilinear" form. Krener's work [13] ensures that this is always possible.

4. High Order Necessary Conditions in the Calculus

Recall that even in the simplest optimization problems involving the minimization of a function $\phi : \mathbb{R}^n \to \mathbb{R}^1$ we have necessary conditions which, when expressed in terms of the Taylor series

$$\phi(z) = \phi_0(z_0) + L_1(z-z_0) + L_2(z-z_0)^{[2]} + \ldots + L_k(z-z_0)^{[k]} + \ldots ,$$

appear as

$$L_1 = 0 \quad \text{and} \quad L_2 \geq 0 ,$$

$$L_3 = 0 \quad \text{and} \quad L_4 \geq 0 \quad \text{on} \ \ker L_2 , \quad\quad (\dagger)$$

$$L_{2i+1} = 0 \quad \text{and} \quad L_{2i+2} \geq 0 \quad \text{on} \ \ker L_2 \cap \ker L_4 \cap \ldots \cap \ker L_{2i}$$

where by $\ker L_i$ we mean the set (not necessarily a linear subspace but a set which is "star-shaped" in the sense that αu belongs if u belongs) on which L_i vanishes. The highest order necessary condition is of order k, where k is the smallest integer such that $\bigcap_{i=1}^{k} \ker L_i$ is trivial. As a consequence, we are justified in thinking of such a problem as being "smooth" only if it is possible to develop the Taylor series as far as is necessary to reveal the basic necessary condition as given by $(\dagger)$. The sequence of numbers

$$\nu_r = \dim\left\{\bigcap_{i=1}^{r} \ker L_i\right\}$$

measures the departure of the singularity of ϕ at z_0 from genericity.

If there are equality constraints of the form

$$0 = \psi(z) = M_1(z-z_0) + M_2(z-z_0)^{[2]} + \ldots$$

then equation (†) must be modified. We have in this case

$$\ker L_1 \supset \ker M_1 \;,$$

$$L_2 \geq 0 \quad \text{on} \quad \ker M_1 \cap \ker M_2 \;,$$

$$\ker L_3 \supset \ker M_1 \cap \ker L_2 \cap \ker M_2 \;,$$

$$L_4 \geq 0 \quad \text{on} \quad \ker M_1 \cap \ker L_2 \cap \ker M_2 \cap \ker M_3 \cap \ker M_4$$

$$\ldots$$

and the analogous set of integers may be defined by

$$\mu_k = \dim \bigcap_{i=1}^{k} \left[(\ker L_i) \cap (\ker M_i) \right] \;.$$

We are justified in thinking of such a problem as being smooth provided there exist enough derivatives of ϕ and ψ to develop these series until the intersection of the kernels is trivial.

5. Singular Optimal Control

Let us consider the problem of minimizing $y(T) = h[x(T)]$ for

$$\dot{x}(t) = f[x(t), t] + u_i(t) g_i[x(t), t] \;;\; x(0) = x_0 \;;\; \phi(x(T)) = 0 \;.$$

Suppose that $u = 0$ is an optimal control. We assume that the problem is smooth in the above sense and would like to describe the necessary conditions in a way which is organized according to the above ideas.

We introduce a definition which gives a means of classifying the degree of singularity of optimal control problems. Consider the expansion for $y(\cdot)$ given by

$$y(t) = y_0(t) + L_1[u(\cdot)] + L_2[u^{[2]}(\cdot, \cdot)] + \ldots \;.$$

We introduce the sets $\ker L_1$, $\ker L_1 \cap \ker L_2$, $\ldots$, $\bigcap_{i=1}^{r} \ker L_i$ and indicate their dimensions by $\nu_1 \geq \nu_2 \geq \ldots \geq \nu_r$. We call the problem *nondegenerate* if $\nu_r = 0$ for some r. Only under this circumstance can the optimal control be unique. If there exist terminal constraints

$$0 = \phi[x(T)] = M_1[x(t)-x_0] + M_2[x(t)-x_0]^{[2]} + \ldots$$

then we want to consider the Volterra series for ϕ as well:

$$\phi[x(t)] = \tilde{M}_1[u(\cdot)] + \tilde{M}_2[u^{[2]}(\cdot,\cdot)] + \ldots .$$

In this case we introduce the modified sequence

$$S_r = \bigcap_{i=1}^{r} \ker L_i \cap \ker M_i$$

and call the problem nondegenerate if $S_r = 0$ for some r. We call $\dim S_r = \nu_r$ the rth *degeneracy index*. A control problem will be said to be *smooth* if f, g_i, h and ϕ are sufficiently smooth to permit one to calculate enough terms in the Volterra series to have $\nu_r = 0$.

The basic necessary conditions are just the same as for finite dimensional problems, namely

$$\ker L_1 \supset \ker M_1 ,$$

$$L_2 \geq 0 \quad \text{on} \quad \ker M_1 \cap \ker M_2 ,$$

$$\ker L_3 \supset \ker M_1 \cap \ker M_2 \cap \ker M_3 \cap \ker L_2 ,$$

$$L_4 \geq 0 \quad \text{on} \quad \ker M_1 \cap \ker M_2 \cap \ker M_3 \cap \ker M_4 \cap \ker L_2$$

$$\ldots .$$

There is one feature of the singular control problem which has no direct analog in terms of function minimization and that is the fact that the necessary conditions may generate conditions which correspond to a lower dimensional set of differential equations. Variational problems for which the jet of the Volterra series which makes $\dim S_r = 0$ has no realization on a manifold of dimension less than the given one are in some sense more typical than others but this theory does not give these problems a special role.

6. The Geometrical Necessary Conditions

The essence of the calculus of variations is to express the necessary conditions in "local form", that is by the Euler-Lagrange differential equation. In the present context one should, then, attempt to understand the meaning of the necessary conditions in terms of the k-jets of f, g, h and ϕ. The insight which is available in this way can be brought out in two ways. On one hand, one can study the way in which the higher order terms of f, y, and so on, enter the expressions for the Volterra kernels (see [5]). An alternative, which has been used much more extensively in the literature is to construct special control variations which generate elements in the range space of particular Volterra kernels without actually constructing the kernels (see [7], [11], [14]).

The structure of the ν sequence is intimately related to the structure of the Lie algebra of vector fields generated by f and $g_1, g_2, \ldots, g_m$. Whenever we distinguish a subset of a Lie algebra which generates that algebra there is a filtration defined on the algebra, the ith subspace being the set of elements which can be expressed in terms of the generating set as a product of i or fewer elements. For our present purposes it is useful to regard $\{f, g_1, g_2, \ldots, g_m\}_{LA}$ as a filtered Lie algebra in a somewhat different way whereby the terms of degree r or less are those which can be expressed as a bracket expression in f and $\alpha g_1, \alpha g_2, \ldots, \alpha g_n$ evaluated at $\alpha = 1$ and which is of degree r or less in α. The distinguished role for f comes about because these problems are local in u but not in t. A key point is that w^k is constructed from terms generating F_k in this filtration. It is then clear that the structure of the necessary conditions is mirrored in the structure of the filtered Lie algebra.

We have then a sequence of subspaces of $L = \{f, g_1, g_2, \ldots, g_n\}_{LA}$. Consider $Ad_f^k(\cdot)$ operator defined inductively by

$$Ad_f^k(g) = \left[f, Ad_f^{k-1}(g)\right] \; ; \; Ad_f^0(g) = g \, ,$$

where

$$[f, g] = \frac{\partial g}{\partial x} f - \frac{\partial f}{\partial x} g$$

denotes the Lie bracket. The L_0 subspace consists of terms of the form $Ad_f^k(g_i)$, $i = 1, 2, \ldots, m$, $k = 0, 1, 2, \ldots$. One sees rather easily that the first degree term in the Volterra series depends only on the structure of L_0. The second degree terms depend only on the structure of L_1, and so on.

The first order necessary condition previously expressed as $\ker L_1 \supset \ker \tilde{M}_1$ now finds its expression in the requirement that there should exist constants α_{ij} such that

$$\left\langle \frac{\partial h}{\partial x}, Ad_f^k g_i \right\rangle = \sum_{l,j} \alpha_{ij} \left\langle \frac{\partial \phi_j}{\partial x}, Ad_f^k g_l \right\rangle .$$

Obviously there is an awkwardness in expressing the conditions locally in that to express the first order conditions locally requires f and g to be C^∞ even though the first order conditions in integrated form required only a continuous first derivative for f and h and even less for g.

Apparently there are no manageable necessary and sufficient conditions on $w(t, \sigma)$ which ensure that

$$\int_0^t \int_0^{\sigma_1} W(\sigma_1, \sigma_2) u^{(2)}(\sigma_1, \sigma_2) d\sigma_1 d\sigma_2 \geq 0 \ .$$

There are, however, a variety of necessary conditions, beginning with the most obvious which is that $w(\sigma_1, \sigma_1) u^{(2)}(\sigma_1, \sigma_1) \geq 0$ for all $u(\sigma_1)$. This means that if we express this kernel as

$$u'(\sigma_1) \tilde{W}(\sigma_1, \sigma_2) u(\sigma_2)$$

then $\tilde{W}(\sigma_1, \sigma_1)$ is a nonnegative definite matrix. It is, however, necessary to translate the vector fields along an integral curve to express $w(\sigma, \sigma)$ for $\sigma \neq 0$ and thus in order to express this locally one needs to make the expansion

$$w(\sigma, \sigma) = w(0, 0) + \sigma w^{(1)}(0, 0) + \frac{\sigma^2}{2} w^{(2)}(0, 0)$$

and use results on the moment problem to get conditions in terms of the vector fields evaluated at $x(0)$.

The results of [11] and [7] concern the condition $\tilde{W}(\sigma, \sigma) \geq 0$ on the kernel of M_1. An alternative approach to higher order conditions is given in Krener [14].

References

[1] Felix Albrecht, *Topics in Control Theory* (Lecture Notes in Mathematics, **63**. Springer-Verlag, Berlin, Heidelberg, New York, 1968).

[2] R.W. Brockett, "System theory on group manifolds and coset spaces", *SIAM J. Control* **10** (1972), 265-284.

[3] Roger W. Brockett, "Nonlinear systems and differential geometry", *Proc. IEEE* **64** (1976), 61-72.

[4] Roger W. Brockett, "Volterra series and geometric control theory", *Automatica, J. IFAC* **12** (1976), 167-176.

[5] Roger W. Brockett, "Functional expansions and higher order necessary conditions in optimal control", *Mathematical Systems Theory* (Lecture Notes in Economics and Mathematical Systems, **131**, 111-121. Springer-Verlag, Berlin, Heidelberg, New York, 1976).

[6] D. Elliott, "Controllable systems driven by white noise" (PhD thesis, University of California, Los Angeles, 1969).

[7] R. Gabasov, "Necessary conditions for optimality of singular control", *Engrg. Cybernetics* (**1968**), no. 5, 28-37 (1969).

[8] Robert Hermann, "On the accessibility problem in control theory", *Internat. Sympos. on Nonlinear Differential Equations and Nonlinear Mechanics*, 325-332 (Academic Press, New York, London, 1963).

[9] H. Hermes and G. Haynes, "On the nonlinear control problem with control appearing linearly", *SIAM J. Control* 1 (1963), 85-108.

[10] Ronald Murray Hirschorn, "Topological semigroups and controllability of bilinear systems" (PhD thesis, Harvard University, 1973).

[11] D.H. Jacobson, "A new necessary condition of optimality for singular control problems", *SIAM J. Control* 7 (1969), 578-595.

[12] Arthur J. Krener, "A generalization of Chow's theorem and the bang-bang theorem to nonlinear control problems", *SIAM J. Control* 12 (1974), 43-52.

[13] Arthur J. Krener, "Linearization and bilinearization of control systems", *Proc. Allerton Conf. Circuit and System Theory*, 1974.

[14] Arthur J. Krener, "The high order maximal principle and its application to singular extremals", *SIAM J. Control Optimization* 15 (1977), 256-293.

[15] James Ting-Ho Lo and Alan S. Willsky, "Estimation for rotational processes with one degree of freedom - Part I: Introduction and continuous-time processes", *IEEE Trans. Automatic Control* AC-20 (1975), 10-21.

[16] C. Lobry, "Quelques aspects qualitatifs de la theories de la commande" (Docteur es Sciences Mathematiques, L'Universite Scientifique et Medicale de Grenoble, 1972).

[17] Héctor J. Sussmann and Velimir Jurdjevic, "Controllability of nonlinear systems", *J. Differential Equations* 12 (1972), 95-116.

[18] Alan S. Willsky and James Ting-Ho Lo, "Estimation for rotational processes with one degree of freedom - Part II: Discrete-time processes", *IEEE Trans. Automatic Control* AC-20 (1975), 22-30.

[19] Alan S. Willsky and James Ting-Ho Lo, "Estimation for rotational processes with one degree of freedom - Part III: Implementation", *IEEE Trans. Automatic Control* AC-20 (1975), 31-33.

NECESSARY CONDITIONS FOR OPTIMAL CONTROL PROBLEMS
WITH DIFFERENTIABLE OR NONDIFFERENTIABLE DATA

Hubert Halkin

Introduction

The aim of this paper is to present a general theory of necessary conditions for optimization problems (mathematical programming, calculus of variations, optimal control theory). Besides problems with differentiable data, I shall also consider problems with nondifferentiable data.

The guiding line of the present paper (and of all my work on necessary conditions during the last 18 years) is the following: the optimality of the solution of an optimization problem can be characterized by the fact that two sets are disjoint; the corresponding necessary condition can be characterized by the fact that two related convex sets are separated[1].

For the benefit of the reader who is not yet familiar with this point of view I will describe below its application to the standard nonlinear differentiable mathematical programming problem. We are given integers $n \geq 1$, $\mu \geq 0$, $m \geq 0$, real-valued functions $\varphi_{-\mu}, \ldots, \varphi_{-1}, \varphi_0, \varphi_1, \ldots, \varphi_m$ defined on the n-dimensional Euclidean space R^n and a convex set $\Omega \subset R^n$. The problem is to find an element

---

This work was partially supported by a National Science Foundation Grant.

[1] The convex sets S_1 and S_2 are separated if there exists a nonzero (continuous) linear functional λ such that
$$\sup_{x \in S_1} \lambda x \leq \inf_{x \in S_2} \lambda x .$$

$x_0 \in \Omega$ which minimizes $\varphi_0[x_0]$ subject to the constraints $\varphi_i[x_0] \leq 0$ for all $i = -\mu, \ldots, -1$ and $\varphi_i[x_0] = 0$ for all $i = 1, \ldots, m$. In other words if A denotes the set

$$\{x : x \in \Omega, \varphi_i[x] \leq 0 \text{ for } i = -\mu, \ldots, -1; \varphi_i[x] = 0 \text{ for } i = 1, \ldots, m\}$$

we want to find an element $x_0 \in A$ such that $\varphi_0[x_0] \leq \varphi_0[x]$ for all $x \in A$. An element x_0 satisfying this condition is said to be optimal.

Let K be the set of all $(\alpha_{-\mu}, \ldots, \alpha_m) \in R^{\mu+m+1}$ such that

$$\alpha_i \leq 0 \text{ for } i = -\mu, \ldots, -1,$$

$$\alpha_0 < \varphi_0[x_0],$$

$$\alpha_i = 0 \text{ for } i = 1, \ldots, m.$$

We denote by φ the function from R^n into $R^{\mu+m+1}$ with components $(\varphi_{-\mu}, \ldots, \varphi_{-1}, \varphi_0, \varphi_1, \ldots, \varphi_m)$ and we let $\varphi[\Omega] \triangleq \{\varphi[x] : x \in \Omega\}$. We remark that $\varphi[\Omega]$ and K are disjoint whenever x_0 is optimal.

Let us assume that the function φ is continuous in a neighborhood of x_0 and admits a derivative $\varphi'[x_0]$ at x_0. The convex set

$$M \triangleq \{\varphi[x_0] + \varphi'[x_0](x-x_0) : x \in \Omega\}$$

is a "linearization" of the set $\varphi[\Omega]$ around the point x_0. An application of the Separation Principle stated below allows us to assert that M and K are separated whenever x_0 is optimal.

SEPARATION PRINCIPLE. If $\Omega \subset R^n$ and $K \subset R^k$ are convex, $x_0 \in \overline{\Omega}$, $\varphi : R^n \to R^k$ is continuous in a neighborhood of x_0 and differentiable at x_0, $\varphi[x_0] \in \overline{K}$, $\varphi[\Omega]$ and K are disjoint then the sets $\{\varphi[x_0] + \varphi'[x_0](x-x_0) : x \in \Omega\}$ and K are separated.

The Separation Principle, a particular case of the results given in the present paper, implies (and is implied by) the Brouwer Fixed Point Theorem (Halkin [15] and [16]). However if the function φ is assumed to be continuously differentiable, then the Separation Principle is an easy consequence of the classical Implicit Function Theorem.

From the Separation Principle we can easily derive the following necessary condition for the nonlinear differentiable mathematical programming problem stated earlier.

CARATHÉODORY-JOHN MULTIPLIER RULE. If x_0 is optimal then there exists a nonzero vector $\lambda = (\lambda_{-\mu}, \ldots, \lambda_m) \in R^{\mu+m+1}$ such that

(1) $\sum_{i=-\mu,\ldots,m} \lambda_i \varphi_i'[x_0](x-x_0) \leq 0$ for all $x \in \Omega$,

(2) $\lambda_i \leq 0$ for $i = -\mu, \ldots, -1, 0$,

(3) $\lambda_i \varphi_i[x_0] = 0$ for $i = -\mu, \ldots, -1$.

An essential feature of this approach to optimization theory is that the analysis is carried out in the *image* space $R^{\mu+m+1} = R^k$ and not in the *domain* space R^n.

In order to obtain interesting necessary conditions in optimal control theory, I shall introduce stronger forms of the Separation Principle: the spaces R^n and R^k will be replaced by appropriate infinite[2] dimensional spaces X and Y, and the assumption of differentiability of the given function will be considerably weakened or even eliminated. Moreover, I shall consider the following form of the Separation Principle which is stable under small perturbations: if the sets K and $\{\varphi[x_0]+\varphi'[x_0](x-x_0) : x \in \Omega\}$ are not separated then there exists an $\eta > 0$ such that for any continuous function $g : \Omega \to Y$ with $|g[x]-\varphi[x]| \leq \eta$ for all $x \in \Omega$ the sets K and $g[\Omega]$ are not disjoint. This form of the Separation Principle has two specific advantages for control theory. First it implies that there exists an $\eta > 0$ such that the sets K and $y^* + \varphi[\Omega]$ are not disjoint for every $y^* \in Y$ with $|y^*| \leq \eta$: in optimal control terminology this means that if a trajectory fails to satisfy the Maximum Principle then a full neighborhood of the terminal state can be reached by the system under consideration (Halkin [6]). Secondly it allows us to say that if a trajectory of a given system fails to satisfy the Maximum Principle then there is a full neighborhood of proximate systems (in an appropriate topology) for which corresponding trajectories will fail to be optimal: this device allows us in effect to give for nonnecessarily relaxed (convexified) optimal control problems a proof of the necessary conditions which is as simple as the proof given in the case of relaxed optimal control problems (Warga [32]).

The culminating point of the present paper is a Maximum Principle for the optimal control of systems described by differential equations in the presence of minimax functionals, bounded-phase constraints and vector-valued objective function when the

---

[2] Replacing the domain space R^n by an infinite dimensional space X is a rather trivial affair. Replacing the image space $R^k = R^{\mu+m+1}$ by an infinite dimensional space Y is a delicate operation.

data is nonlinear, nonconvex and nondifferentiable.

1. Differentiable Separation Principle between Banach Spaces

In Proposition 1.1 of the present chapter I shall give a form of the Separation Principle which is sufficient to prove necessary conditions for a wide class of optimal control problems for systems described by nonlinear differential equations when the data is assumed to satisfy some differentiability conditions. This class of optimal control problems is large enough to include vector-valued objective functions, minimax functionals and bounded phase constraints.

Proposition 1.1 is a Separation Principle between Banach spaces. The reader should be aware that a Separation Principle between finite dimensional Euclidean spaces would be sufficient for a wide subclass of the above optimal control problems, including problems with vector-valued objective functions, equality and inequality constraints on the state variables at a finite number of instants. The specific interest of a Separation Principle between Banach spaces can be found in its applications to optimal control problems with bounded phase constraints and minimax functionals.

NOTATION. If X and Y are Banach spaces I shall denote by $B(X, Y)$ the set of all continuous linear functions from X into Y. If $Y = R^1$, the real line, I shall use the notation X^* instead of $B(X, R^1)$. The closure of a subset Ω of a Banach space X will be denoted by $\overline{\Omega}$. If $g : \Omega \to Y$ then $g[\Omega]$ will denote the set $\{g[x] : x \in \Omega\}$.

DEFINITION. Two nonempty convex subsets M_1 and M_2 of a Banach space Y are said to be *separated* if there exists a nonzero element $\lambda \in Y^*$ such that
$$\sup_{y \in M_1} \lambda y \leq \inf_{y \in M_2} \lambda y .$$

DEFINITION. A convex subset K of a Banach space Y is said to be *substantial* if K has a nonempty relative interior with respect to a closed affine subspace of Y of finite codimension, that is, if there exists an element $y^* \in K$, a real number $\varepsilon > 0$, an integer $m \geq 0$ and a continuous affine function $p : Y \to R^m$ such that

(1) $K \subset \{y : y \in Y, p[y] = 0\}$,

(2) $\{y : y \in Y, p[y] = 0, |y-y^*| < \varepsilon\} \subset K$.

REMARK. A nonempty convex subset of a finite dimensional Euclidean space is always substantial.

PROPOSITION 1.1. *If X and Y are Banach spaces, $\Omega \subset X$ and $K \subset Y$ are convex, K is substantial, $f : X \to Y$ is Fréchet differentiable at $x_0 \in \overline{\Omega}$, $f[x_0] \in \overline{K}$, the sets $\{f[x_0]+f'[x_0](x-x_0) : x \in \Omega\}$ and K are not separated, then* there exists an $\eta > 0$ such that for any continuous function $g : \Omega \to Y$ with $|g[x]-f[x]| \leq \eta$ for all $x \in \Omega$ the sets $g[\Omega]$ and K are not disjoint.

Before proving Proposition 1.1, I will need to state and prove Lemma 1.1. The Kakutani-Fan Fixed Point Theorem is the critical element of the proof of Lemma 1.1.

NOTATION. If X is a Banach space, $x_0 \in X$, $S \subset X$ and $\varepsilon > 0$ then $N(x_0, \varepsilon)$ will denote the set $\{x : x \in X, |x-x_0| < \varepsilon\}$ whereas $N(S, \varepsilon)$ will denote the set $\bigcup_{x \in S} N(x, \varepsilon)$. If S is a nonempty bounded subset of X then $|S|$ will denote the number $\sup_{x \in S} |x|$.

LEMMA 1.1. *If X is a Banach space, Y_+ is a finite dimensional Euclidean space, $S \subset X$ is compact and convex, $\sigma > 0$, $g_+ : S \to Y_+$ is continuous, $a \in B(X, Y_+)$, $N(0, \sigma) \subset aS$ and $|g_+[x]-ax| \leq \sigma$ for all $x \in S$ then $0 \in g_+[S]$.*

Proof of Lemma 1.1. We shall assume that $0 \notin g_+[S]$ and show that we are led to a contradiction. If $0 \notin g_+[S]$ then $N(0, \delta) \cap g_+[S] = \emptyset$ for some $\delta > 0$. Let[3] $E[x, x^*] = \dfrac{g_+[x]}{|g_+[x]|} \cdot (ax^* - g_+[x])$ for all $x, x^* \in S$. We have $|E[x, x]| \leq \sigma$ for all $x \in S$. For every $x \in S$ let $F[x]$ be a subset of S defined by $F[x] = \{x^* : x^* \in S, E[x, x^*] \leq -\sigma-\delta\}$. For every $x \in S$ the set $F[x]$ is compact, convex and nonempty (since for every $x \in S$ there exists an $\bar{x} \in S$ such that $a\bar{x} = -\sigma \dfrac{g_+[x]}{|g_+[x]|}$ and hence such that $E[x, \bar{x}] \leq -\sigma-\delta$). Moreover we have $x^* \in F[x]$ whenever $x^* = \lim_{i \to \infty} x_i^*$, $x = \lim_{i \to \infty} x_i$ and $x_i^* \in F[x_i]$ for every $i \in \{1, 2, \ldots\}$ (since the mapping $(x, x^*) \to E[x, x^*] : S \times S \to R^1$ is continuous). It then follows that the mapping $x \to F[x]$ is upper semicontinuous. From the Kakutani-Fan Theorem we know that we have $x^+ \in F[x^+]$ for some $x^+ \in S$ and hence $E[x^+, x^+] \leq -\sigma-\delta$. We have then $|E[x^+, x^+]| \geq \sigma+\delta$ which contradicts the relation $|E[x, x]| \leq \sigma$ for all $x \in S$. This concludes the proof of Lemma 1.1.

Proof[4] of Proposition 1.1. We may assume without loss of generality that $x_0 = 0$ and that $f[0] = 0$. We may also assume that $Y = Y_- \times Y_+$ and that $K = K_- \times \{0\}$ where K_- is a convex subset with nonempty interior of a Banach space Y_- and where Y_+ is a finite dimensional Euclidean space. We have then $0 \in \overline{K_-}$. Let $f_- : X \to Y_-$ and $f_+ : X \to Y_+$ such that $f[x] = (f_-[x], f_+[x])$ for all $x \in X$. Let $f'_-[0] \in B(X, Y_-)$ and $f'_+[0] \in B(X, Y_+)$ such that $f'[0]x = (f'_-[0]x, f'_+[0]x)$ for all

---

[3] The scalar product of two vectors u and v in a finite dimensional Euclidean space is denoted by $u \cdot v$. Please note that everywhere else in this paper I am avoiding scalar products and using elements of the dual instead.

[4] The proof of Proposition 1.1 can be considerably simplified when the space Y is assumed to be finite dimensional. See Comment 1.8.

$x \in X$.

Let $\Omega_+ = \{x : x \in \Omega, f'_-[0]x \in \text{int } K_-\}$. Since the sets $K_- \times \{0\}$ and $\{(f'_-[0]x, f'_+[0]x) : x \in \Omega\}$ are not separated it follows from Hahn-Banach Theorem[5] that Ω_+ is not empty and that $0 \in \text{int } f'_+[0]\Omega_+$. Since Ω_+ is not empty, $0 \in \overline{\Omega}_+$, int K_- is open, and $0 \in \overline{K}_-$ we have $0 \in \overline{\Omega}_+$.

Since Y_+ is finite dimensional there exists a finite subset T of Ω_+ and an $\varepsilon_1 > 0$ such that[6] $N(0, \varepsilon_1) \subset f'_+[0]\text{co } T$. We have $f'_-[0]x \in \text{int } K_-$ for all $x \in \text{co } T$. Since T is finite there exists some $\varepsilon_2 > 0$ such that $N(f'_-[0]x, \varepsilon_2) \subset K_-$ for all $x \in \text{co } T$. Let $\varepsilon = \min\{\varepsilon_1, \varepsilon_2\}$. Since Ω_+ is convex, $0 \in \overline{\Omega}_+$ and $0 \in \overline{K}_-$ we can find a sequence $T_1, T_2, \ldots$ of finite subsets of Ω_+ such that for all $i \in \{1, 2, \ldots\}$ we have

(1) $|T_i| \leq \frac{1}{i}|T|$,

(2) $N\left(0, \frac{\varepsilon}{2i}\right) \subset f'_+[0]\text{co } T_i$,

(3) $N\left(f'_-[0]\text{co } T_i, \frac{\varepsilon}{2i}\right) \subset K_-$.

Let $j \in \{1, 2, \ldots\}$ be such that for all $x \in \text{co } T_j$ we have $|f'_+[0]x - f_+[x]|$ and $|f'_-[0]x - f_-[x]| \leq \frac{\varepsilon}{4j}$. Let $S = \text{co } T_j$ and let $\eta = \frac{\varepsilon}{4j}$. Let g be a continuous function from Ω into Y such that $|g[x] - f[x]| \leq \eta$ for all $x \in \Omega$. Let $g_+ : \Omega \to Y_+$ and $g_- : \Omega \to Y_-$ such that $g[x] = (g_-[x], g_+[x])$ for all $x \in \Omega$. We know from relation (3) that for all $x \in S \subset \Omega$ we have $g_-[x] \in K_-$ since $|g_-[x] - f'_-[0]x| \leq |g_-[x] - f_-[x]| + |f_-[x] - f'_-[0]x| \leq \frac{\varepsilon}{4j} + \frac{\varepsilon}{4j} = \frac{\varepsilon}{2j}$. Moreover we know from Lemma 1.1 (with $\sigma = \frac{\varepsilon}{2j} = 2\eta$) that $0 \in g_+[S]$ since $N\left(0, \frac{\varepsilon}{2j}\right) \subset f'_+[0]S$ and since for all $x \in S$ we have

$$|g_+[x] - f'_+[0]x| \leq |g_+[x] - f_+[x]| + |f_+[x] - f'_+[0]x| \leq \frac{\varepsilon}{4j} + \frac{\varepsilon}{4j} = \frac{\varepsilon}{2j}.$$

For some $x \in S$ we have $g_+[x] = 0$ and for all $x \in S$ we have $g_-[x] \in K_-$; it does follow that the sets $g[S]$ and $K = K_- \times \{0\}$ are not disjoint and *a fortiori* that

---
[5] Indeed if $\Omega_+ = \emptyset$ then K_- and $f'_-[0]\Omega$ are separated by some nonzero $\lambda_- \in (Y_-)^*$ and hence $K_- \times \{0\}$ and $(f'_-[0], f'_+[0])\Omega$ are separated by $(\lambda_-, 0)$. If $\Omega_+ \neq \emptyset$ and $0 \notin \text{int } f'_+[0]\Omega_+$ then for some nonzero $\lambda_+ \in (Y_+)^*$ we have $\lambda_+ y_+ \leq 0$ for all $y_+ \in f'_+[0]\Omega_+$; the sets $(f'_-[0], f'_+[0])\Omega$ and $K^* = \{(y_-, y_+) : y_- \in \text{int } K_-, \lambda_+ y_+ > 0\}$ are disjoint and hence separated since int $K^* \neq \emptyset$, and hence the sets $(f'_-[0], f'_+[0])\Omega$ and $K_- \times \{0\}$ are separated.
[6] The convex hull of the set S is denoted by co S.

the sets $g[\Omega]$ and K are not disjoint. This concludes the proof of Proposition 1.1.

COMMENT 1.1. The reader should be aware that the results of the present chapter would still be valid (and not a single word of the proof needs to be changed) if we would assume that, instead of admitting a Fréchet derivative $f'[x_0]$ at x_0, the function f admits a *finite Gâteaux* derivative $f'[x_0]$ at x_0, that is, if we would assume that the restriction of $f'[x_0]$ to any finite dimensional affine subspace of X passing through x_0 is the derivative of the restriction of f to that subspace. Moreover this requirement can be limited to the affine subspaces of of dimensions not exceeding the codimension of the closed affine span of K.

If an optimization problem includes composition of functions (this is the case of the control problem considered in Chapter 2 of the present paper) the type of derivative adopted should be appropriate not only for Proposition 1.1 but also for a chain rule. The reader should be aware that the finite Gâteaux derivative *does not* obey the chain rule but that the Fréchet derivative does. The Hadamard derivative, which is stronger than the finite Gâteaux derivative and weaker than the Fréchet derivative, is the weakest derivative satisfying the chain rule. Although the Hadamard derivative would give the strongest overall results in optimization theory, I have chosen to keep using the (simpler and better known) Fréchet derivative in my papers on differentiable optimal control theory as long as no interesting and concrete control problem appears with a set of data which is Hadamard but not Fréchet differentiable. For the study of Hadamard derivatives see Yamamuro [34].

COMMENT 1.2. The results of the present chapter can be extended to the case of a linear space X and of a linear topological space Y. In that case the words "S is compact and convex" appearing in Lemma 1.1 should be replaced by the words "S is the convex hull of a finite set". This weaker version of Lemma 1.1 would also be sufficient to prove Proposition 1.1 in the case of Banach spaces X and Y. Please note that the Kakutani Fixed Point Theorem is sufficient for the proof of the weaker version of Lemma 1.1.

COMMENT 1.3. Proposition 1.1 remains true if, instead of assuming the continuity of g, we assume only the continuity of g_+ as defined in the proof of Proposition 1.1.

COMMENT 1.4. Proposition 1.1 is false without the assumption "K is substantial". Indeed let $X = Y$ and let f be the identity mapping of X into $X = Y$. Two disjoint nonempty convex subsets of a Banach space are not always separated. The Hahn-Banach Theorem, for instance, requires that one of the sets has a non-empty interior.

COMMENT 1.5. If the function f is assumed to be continuously differentiable then it is relatively easy to prove that the sets $f[\Omega]$ and K are not disjoint (Implicit Function Theorem).

COMMENT 1.6. If the set K is assumed to have a nonempty interior then the proof of Proposition 1.1 is very easy.

COMMENT 1.7 (Multiplier Rule for Differentiable Mathematical Programming Problem). I shall give below a multiplier rule for a mathematical programming problem over a Banach space in the presence of a vector-valued objective function, a finite number of equality constraints and an infinite number of inequality constraints.

We are given Banach spaces X, Y_1 and Y_2, a finite-dimensional Euclidean space Y_3, a convex set $\Omega \subset X$, convex cones $K_1 \subset Y_1$ and $K_2 \subset Y_2$ with $0 \notin K_1$ and functions $f_i : X \to Y_i$ for $i = 1, 2,$ and 3. Let

$$\tilde{A} = \{x : x \in \Omega, f_2[x] \in K_2, f_3[x] = 0\} .$$

The problem is to find an element $x_0 \in \tilde{A}$ such that we have $f_1[x] - f_1[x_0] \notin K_1$ for all $x \in \tilde{A}$. We assume that K_1 and K_2 have nonempty interiors, that f_3 is continuous in a neighborhood of x_0 and that f_i is Fréchet differentiable at x_0 for $i = 1, 2$ and 3. Let $K_1^+ = \{f_1[x_0] + y_1 : y_1 \in K_1\}$ and let $K = K_1^+ \times K_2 \times \{0\}$. From Proposition 1.1 and Comment 1.3 we see immediately that if x_0 is optimal then there exists a nonzero element $(\lambda_1, \lambda_2, \lambda_3) \in Y_1^* \times Y_2^* \times Y_3^*$ such that

(1) $\sum_{i=1,2,3} \lambda_i f_i'[x_0](x-x_0) \leq 0$ for all $x \in \Omega$,

(2) $\lambda_1 y_1 \geq 0$ for all $y_1 \in K_1$,

(3) $\lambda_2 y_2 \geq 0$ for all $y_2 \in K_2$,

(4) $\lambda_2 f_2[x_0] = 0$.

The above result is a generalization to Banach spaces of the multiplier rule given in Halkin [12]: no continuity assumption is made for the functions describing the objective and inequality constraints.

COMMENT 1.8 (Interior Mapping Theorem). When the space Y is assumed to be finite dimensional it is possible to give a simpler proof of Proposition 1.1. This simpler proof, which is given below, is based on the following Interior Mapping Theorem:

LEMMA 1.2. *If X is a Banach space, Y is a finite dimensional Euclidean* space, $\Omega \subset X$ is convex, $f : X \to Y$ is differentiable[7] at $x_0 \in \overline{\Omega}$, $0 \in \text{int}\{f'[x_0](x-x_0) : x \in \Omega\}$ then there exists an $\eta > 0$ such that for any continuous function $g : \Omega \to Y$ with $|g[x]-f[x]| \leq \eta$ for all $x \in \Omega$ we have $N(f[x_0], \eta) \subset g[\Omega]$.

Proof of Lemma 1.2. We may assume without loss of generality that $x_0 = 0$ and $f[0] = 0$. There exists an $\varepsilon > 0$ and a finite set $S \subset \Omega$ such that[8] $N(0, \varepsilon) \subset f'[0]\text{co } S$. Since Ω is convex and since $0 \in \overline{\Omega}$ then for every $i \in \{1, 2, \ldots\}$ there exists a finite set $S_i \subset \Omega$ such that $|S_i| \leq \frac{2}{i}|S|$ and $N\left(0, \frac{\varepsilon}{i}\right) \subset f'[0]\text{co } S_i$. There exists a $j \in \{1, 2, \ldots\}$ such that $|f'[0]x-f[x]| \leq \frac{\varepsilon}{3j}$ for all $x \in \text{co } S_j$. The number $\eta = \frac{\varepsilon}{3j}$ satisfies the required condition. Indeed, let y be an arbitrary element of Y with $|y| \leq \frac{\varepsilon}{3j}$ and let g be a continuous function from Ω into Y such that $|g[x]-f[x]| \leq \frac{\varepsilon}{3j}$ for all $x \in \Omega$. For all $x \in \text{co } S_j$ we have then $|f'[0]x-g[x]+y| \leq |f'[0]x-f[x]| + |f[x]-g[x]| + |y| \leq \frac{\varepsilon}{j}$. Since $N\left(0, \frac{\varepsilon}{j}\right) \subset f'[0]\text{co } S_j$ we obtain, from Lemma 1.1, that $0 \in \{g[x]-y : x \in \text{co } S_j\}$ and hence $y \in g[\text{co } S_j] \subset g[\Omega]$. This concludes the proof of Lemma 1.2.

Proof[9] of Proposition 1.1. There is no loss of generality by assuming that $x_0 = 0$ and $f[0] = 0$. Let $\varphi : X \times Y \to Y$ be defined by $\varphi[x, y] = f[x] - y$. We have $\varphi[0, 0] = 0$ and $\varphi'[0, 0] = (f'[0], -I)$ where I denotes the identity mapping from Y into Y. Since the sets $f'[0]\Omega$ and K are not separated we have $0 \in \text{int}\{f'[0]x-y : x \in \Omega, y \in K\}$. We have also $(0, 0) \in \overline{\Omega \times K}$. From Lemma 1.2 we know that there exists an $\eta > 0$ such that for all continuous functions $\psi : \Omega \times K \to Y$ with $|\psi[x, y]-\varphi[x, y]| \leq \eta$ for all $(x, y) \in \Omega \times K$ we have $0 \in \psi[\Omega, K]$. In particular, if $\psi[x, y] = g[x] - y$ for some continuous function $g : \Omega \to Y$ with $|g[x]-f[x]| \leq \eta$ for all $x \in \Omega$ then we have $0 \in g[\Omega] - K$ and hence $g[\Omega] \cap K \neq \emptyset$. This concludes the proof of Proposition 1.1.

The Interior Mapping Theorem (Lemma 1.2) underlines, in my opinion, what is really the essential element of the classical Implicit and Inverse Function Theorems.

---

[7] It is sufficient to assume that $f'[x_0]$ is the finite Gâteaux derivative of f at x_0, that is, that the restriction of $f'[x_0]$ to any finite dimensional affine subspace of X passing through x_0 is the Fréchet derivative of the restriction of f to that subspace. Moreover this requirement can be limited to the affine subspaces of X of dimensions not exceeding the dimension of Y.
[8] The convex hull of the set S is denoted by $\text{co } S$.
[9] When Y has finite dimension. The general proof has been given earlier.

Moreover the Interior Mapping Theorem is often a more convenient tool since it does not depend on the assumptions of "interiority" (that is, $x_0 \in \text{int } \Omega$) and continuous differentiability (or strong differentiability) required for the Implicit and Inverse Function Theorems. Although the relevance to optimization theory of the Interior Mapping Theorem (or at least its interior-continuously differentiable version) was recognized more than fifty years ago by Carathéodory [2], it has been unfortunately ignored by most recent students of optimization theory.

Lemma 1.2, like Proposition 1.1, implies and is implied by the Brouwer Fixed Point Theorem (Halkin [15] and [16]).

2. Separation Principle for Differentiable Control Systems

In Proposition 2.1 of the present chapter I shall state and prove a Separation Principle for a control system described by a family of nonlinear differential equations. Proposition 2.1 is a stylized form of the Maximum Principle for control problems with minimax functionals, bounded-phase constraints and vector-valued objective function when the data is assumed to satisfy some differentiability conditions. In Chapter 3 I will describe a few concrete interpretations of Proposition 2.1.

The state of the control system will be represented by an element x of a finite dimensional Euclidean space X. I shall denote by $C(X)$ the space of all continuous functions from $[0, 1]$ into X. I am given a family Γ of functions from $X \times [0, 1]$ into X. Each element $f \in \Gamma$ describes a dynamically feasible mode of operation of the control system. An element $\varphi \in C(X)$ is a solution for $f \in \Gamma$ if

$$\varphi[t] - \varphi[0] = \int_0^t f[\varphi[\tau], \tau] d\tau \text{ for every } t \in [0, 1].$$ I shall denote by A the set of all $\varphi \in C(X)$ such that φ is a solution for some $f \in \Gamma$. I am given a Banach space[10] Z, a function g from $C(X)$ into Z and a substantial (see definition in Chapter 1) convex set $K \subset Z$. I shall assume that the sets K and $g[A] \triangleq \{g[\varphi] : \varphi \in A\}$ are disjoint and that for some $\varphi_0 \in A$ we have $g[\varphi_0] \in \overline{K}$. I shall denote by f_0 the element of Γ such that φ_0 is a solution for f_0. The aim of the present chapter is to prove in Proposition 2.1 the existence of a nonzero linear functional separating K and some approximation of $g[A]$ given in terms of derivatives of f_0 and g at φ_0.

NOTATIONS. In this chapter, I shall not use explicitly the scalar product of two

---

[10] If the space Z is assumed to be a finite-dimensional Euclidean space then Proposition 2.1 will not apply to problems with bounded phase constraints or minimax functionals; however, it will still apply to problems with vector-valued objective functions, equality and inequality constraints on the state variables at a finite number of instants.

elements of a finite dimensional Euclidean space, but I shall work instead with elements of the dual. On $C(X)$ I shall use the norm $|\cdot|$ defined by the relation $|\varphi| = \sup_{t \in [0,1]} |\varphi[t]|$. The space $C(X)$ endowed with the norm $|\cdot|$ is a Banach space. The space of all continuous linear mappings from $C(X)$ into Z will be denoted by $B(C(X), Z)$. If $\omega \in B(C(X), Z)$ and $q \in Z^*$ then $q\omega \in (C(X))^*$ and $q\omega$ can be identified with an X^*-valued Borel measure[11] representing it according to Riesz Theorem; that is, for every $\varphi \in C(X)$ and $q\omega \in (C(X))^*$ I shall write $q\omega\varphi = \int_0^1 q\omega[dt]\varphi[t] \in R^1$. If T is a Borel subset of $[0, 1]$ then $q\omega(T) \in X^*$ will denote the $q\omega$ measure of the set T. For instance if $t \in [0, 1]$ then $q\omega([t, 1]) \in X^*$ will denote the $q\omega$ measure of the set $[t, 1]$.

Let $\rho > 0$, $\sigma : [0, 1] \to [0, \infty)$ be integrable, $\hat{\varphi} \in C(X)$ and $h : X \times [0, 1] \to X_+$, where X_+ is a finite dimensional normed linear space. I shall say that h is of class $\Xi(\rho, \sigma)$ around $\hat{\varphi}$ if

(1) for every $x \in X$ the function $t \to h[x, t]$ is measurable,

(2) $|h[x, t]| \leq \sigma[t]$ whenever $t \in [0, 1]$ and $|x-\hat{\varphi}[t]| \leq \rho$,

(3) $|h[x_2, t]-h[x_1, t]| \leq \sigma[t]|x_2-x_1|$ whenever $t \in [0, 1]$, $|x_1-\hat{\varphi}[t]|$, $|x_2-\hat{\varphi}[t]| \leq \rho$.

I shall say that h is of class Ξ around $\hat{\varphi}$ if h is of class $\Xi(\rho, \sigma)$ around $\hat{\varphi}$ for some $\rho > 0$ and some integrable function $\sigma : [0, 1] \to [0, \infty)$.

ASSUMPTIONS. H1 The family Γ is convex-under-switching, that is, for every f_1 and $f_2 \in \Gamma$ and every $\tau \in (0, 1]$ the function $f : X \times [0, 1] \to X$ defined by

$$f[x, t] = f_1[x, t] \text{ if } t \in [0, \tau)$$
$$= f_2[x, t] \text{ if } t \in [\tau, 1]$$

will also belong to Γ.

H2 Every $f \in \Gamma$ is of class Ξ around φ_0.

H3 The function g is uniformly continuous in a neighborhood of φ_0 and admits a Fréchet derivative $\omega \in B(C(X), Z)$ at φ_0.

---

[11] The reader who is not interested in problems with bounded phase constraints or minimax functionals can assume that all X^*-valued measures encountered in this chapter have finite support, that is, that there exist a finite subset $\{t_1, t_2, \ldots, t_k\}$ of $[0, 1]$ and a finite subset $\{p_1, p_2, \ldots, p_k\}$ of X^* such that for any (measurable) subset T of $[0, 1]$ we have $q\omega(T) = \sum_{i \in \{1,\ldots,k\}, t_i \in T} p_i$ and such that for any $\varphi \in C(X)$ we have $q\omega\varphi = \sum_{i \in \{1,\ldots,k\}} p_i \varphi[t_i]$.

H4 An integrable function $E : [0, 1] \to B(X, X)$ is given such that for every $t \in [0, 1]$ the function $x \to f_0[x, t]$ admits $E[t]$ as its derivative at the point $\varphi_0[t]$.

PROPOSITION 2.1. *If* $g[\varphi_0] \in \overline{K}$ *and* $g[A] \cap K = \emptyset$ *then there exists a nonzero element* $q \in Z^*$ *such that*

(1) $q(z-g[\varphi_0]) \geq 0$ *for all* $z \in K$,

(2) *the unique bounded integrable solution* $p_q : [0, 1] \to X^*$ *of the integral equation*

$$p_q[t] = \int_t^1 p_q[\tau]E[\tau]d\tau + q\omega([t, 1]) \quad \text{for every} \quad t \in [0, 1]$$

will satisfy the conditions

(3) $p_q[0] = 0$,

(4) $\int_0^1 p_q[t](f_0[\varphi_0[t], t] - f[\varphi_0[t], t])dt \geq 0$ *for all* $f \in \Gamma$.

REMARK 1. We see immediately that p_q will be a left continuous function of bounded variation.

REMARK 2. From Assumption H1 and relation (4) it follows that

(4)* $$\int_{t_0}^{t_1} p_q[t](f_0[\varphi_0[t], t] - f[\varphi_0[t], t])dt \geq 0$$

for all $f \in \Gamma$ and all $t_0 \leq t_1$ in $[0, 1]$.

PRELIMINARY DEFINITION. Let S_1 be a given set, let S_2 be a given linear space and let $h_0, h_1, \ldots, h_l$ be given functions from $S_1 \times [0, 1]$ into S_2. Let Y be an l-dimensional space with norm $|(y_1, \ldots, y_l)| = \sum_{i=1,\ldots,l} |y_i|$. Let $P = \{y = (y_1, \ldots, y_l) : y \in Y, y_i \geq 0 \text{ for } i = 1, \ldots, l \text{ and } |y| \leq 1\}$.

(A) For every $y \in Y$, I define a function $h_y : S_1 \times [0, 1]$ into S_2 by the relation $h_y = h_0 + \sum_{i=1,\ldots,l} y_i(h_i - h_0)$. The function h_y is called the y affine combination of the functions $h_0, h_1, \ldots, h_l$.

(B) For every $y \in P$, and every $j \in \{1, 2, \ldots\}$ I define a function $h_{y,j}$

from $S_1 \times [0, 1]$ into S_2 by the relation[12]

$$h_{y,j}[s, t] = h_i[s, t]$$

if

$$jt \in \bigcup_{n=1,\ldots,j} \left[n - \sum_{k=i,\ldots,l} y_k,\ n - \sum_{k=i+1,\ldots,l} y_k \right)$$

and by the relation

$$h_{y,j}[s, t] = h_0[s, t]$$

otherwise. The function $h_{y,j}$ is called the y, j switching combination of the functions $h_0, h_1, \ldots, h_l$.

The preceding concepts will also be used when S_1 contains a single point (that is, when the given functions depend only on the variable $t \in [0, 1]$).

Proof of Proposition 2.1. In Chapter 4, I shall prove[13] (using the fact that the convex set K is substantial) that there exists a weak* compact set $Q \subset Z^*$ such that $0 \notin Q$ and such that for every nonzero element $q \in Z^*$ with $q(z - g[\varphi_0]) \geq 0$ for all $z \in K$ we have $\eta q \in Q$ for some $\eta > 0$. For every subset Σ of Γ let $T(\Sigma)$ be the set of all $q \in Q$ such that relations (1), (2) and (3) hold and such that relation (4) holds for all $f \in \Sigma$. I shall prove Proposition 2.1 by showing that for every *finite* subset Σ of Γ the set $T(\Sigma)$ is nonempty and weak* compact.

Let $\Sigma = \{f_1, \ldots, f_l\}$ be an arbitrary subset of Γ. For any $y \in Y$ let $f_y : X \times [0, 1] \to X$ be the y affine combination of the functions $f_0, f_1, \ldots, f_l$. For any $y \in P$ and $j \in \{1, 2, \ldots\}$ let $f_{y,j} : X \times [0, 1] \to X$ be the y, j switching combination of the functions $f_0, f_1, \ldots, f_l$. From Assumption H1 we know that $f_{y,j} \in \Gamma$ for every $y \in P$ and every $j \in \{1, 2, \ldots\}$. In general we do not have $f_y \in \Gamma$. In Chapter 5 I shall prove the existence of a number $\alpha > 0$ such that the sets

$$\theta = \{(x, y) : x \in X, y \in Y, |x| \text{ and } |y| \leq \alpha\}$$

and

$$\Delta = \{(x, y) : x \in X, y \in P, |x| \text{ and } |y| \leq \alpha\}$$

---

[12] I use the convention $\sum_{k=l+1, l} y_k = 0$.

[13] The reader who is only interested in the case of a finite dimensional space Z should know that for that case (1) a nonempty convex set $K \subset Z$ is always substantial, (2) the weak* topology on Z^* is identical with the Euclidean topology on Z^* and (3) the set $Q = \{q : q \in Z^*, |q| = 1\}$ satisfies the required properties.

have the following properties:

(5) for every $(x, y) \in \theta$ there exists a unique element $\varphi_{x,y} \in C(X)$ such that

$$\varphi_{x,y}[t] = \varphi_0[0] + x + \int_0^t f_y\bigl[\varphi_{x,y}[\tau], \tau\bigr] d\tau$$

for every $t \in [0, 1]$.

Moreover, the mapping $(x, y) \to \varphi_{x,y}$ is continuous.

(6) For every $(x, y) \in \Delta$ and every $j \in \{1, 2, \ldots\}$ there exists a unique element $\varphi_{x,y,j} \in C(X)$ such that

$$\varphi_{x,y,j}[t] = \varphi_0[0] + x + \int_0^t f_{y,j}\bigl[\varphi_{x,y,j}[\tau], \tau\bigr] d\tau$$

for every $t \in [0, 1]$.

Moreover, for every $j \in \{1, 2, \ldots\}$ the mapping $(x, y) \to \varphi_{x,y,j}$ is continuous.

(7) The mappings $(x, y) \to \varphi_{x,y,j}$ converge uniformly on Δ to the mapping $(x, y) \to \varphi_{x,y}$ as j tends to $+\infty$.

Let D be the integrable function from $[0, 1]$ into $B(Y, X)$ defined by

$$D[t]y = \sum_{i=1,\ldots,l} y_i \bigl(f_i[\varphi_0[t], t] - f_0[\varphi_0[t], t]\bigr)$$

for all $t \in [0, 1]$ and all $y \in Y$.

In Chapter 6, I shall prove that there exists a unique element $H \in B(X \times Y, C(X))$ such that

(8) $$\bigl(H(x, y)\bigr)[t] = x + \left(\int_0^t D[\tau] d\tau\right) y + \int_0^t E[\tau]\bigl(H(x, y)\bigr)[\tau] d\tau$$

for every $x \in X$, $y \in Y$ and $t \in [0, 1]$. A routine computation shows that H is the Fréchet derivative of $(x, y) \to \varphi_{x,y}$ at $(0, 0)$.

We now define the mapping $\psi : \theta \to Z$ by $\psi[x, y] = g[\varphi_{x,y}]$ and for every $j \in \{1, 2, \ldots\}$ we define the mapping $\psi_j : \Delta \to Z$ by $\psi_j[x, y] = g[\varphi_{x,y,j}]$. We may assume without loss of generality that the number $\alpha > 0$ is small enough such that for every $(x, y) \in \theta$ the element $\varphi_{x,y}$, (respectively for every $(x, y) \in \Delta$ and every $j \in \{1, 2, \ldots\}$ the element $\varphi_{x,y,j}$) will belong to the neighborhood of φ_0 on which we have assumed that the function g is uniformly continuous.

The mapping ψ is continuous on θ. For every $j \in \{1, 2, \ldots\}$ the mapping ψ_j is continuous on Δ. Since g is uniformly continuous and since the functions $(x, y) \to \varphi_{x,y,j}$, $j \in \{1, 2, \ldots\}$ converge uniformly on Δ to the function $(x, y) \to \varphi_{x,y}$ it follows that the mappings $\psi_1, \psi_2, \ldots$ converge uniformly to ψ on Δ. The mapping $\psi : \theta \to Z$ admits the element $\omega H \in B(X \times Y, Z)$ as Fréchet derivative at $(0, 0)$ according to the chain rule for Fréchet derivatives.

Since $\psi[0, 0] = g[\varphi_0] \in \overline{K}$ and since for every $j \in \{1, 2, \ldots\}$ we have $\psi_j[\Delta] \subset g[A]$ and hence $\psi_j[\Delta] \cap K = \emptyset$ it follows from Proposition 1.1 that the sets K and $\{g[\varphi_0] + \omega H(x, y) : (x, y) \in \Delta\}$ are separated, that is that the set $F \stackrel{\Delta}{=} \{q \in Q : q(z-g[\varphi_0]) \geq 0 \text{ for all } z \in K; q\omega H(x, y) \leq 0 \text{ for all } (x, y) \in \Delta\}$ is not empty. Moreover the set F is weak* compact (since F is a weak* closed subset of the weak* compact set Q). We shall conclude the proof of Proposition 2.1 by showing that $F = T(\Sigma)$.

In Chapter 6 I shall prove that for every $q \in Q$ there exists a unique bounded integrable function $p_q : [0, 1] \to X^*$ satisfying relation (2) and that for any $(x, y) \in X \times Y$ we have

(9) $$q\omega H(x, y) = p_q[0]x + \left(\int_0^1 p_q[\tau]D[\tau]d\tau\right)y .$$

From relation (9) we see that we have $q\omega H(x, y) \leq 0$ for all $(x, y) \in \Delta$ if and only if

(10) $$p_q[0]x \leq 0 \text{ for all } x \in X \text{ with } |x| \leq \alpha$$

and

(11) $$\left(\int_0^1 p_q[\tau]D[\tau]d\tau\right)y \leq 0 \text{ for all } y \in P .$$

Relation (10) is equivalent to relation (3). From the definition of D and P we see that relation (11) is equivalent to relation (4) for all $f \in \Sigma$. Since relations (1) and (2) are integral parts of the definitions of both F and $T(\Sigma)$, we have thus shown that $F = T(\Sigma)$. This concludes the proof of Proposition 2.1.

COMMENT 2.1. Assumption H3 is stronger than necessary. We need only to know that ωH is the finite Gâteaux derivative of $(x, y) \to g[\varphi_{x,y}]$ at $(0, 0)$; see Comment 1.1. For instance it would be sufficient to assume that ω is the Hadamard derivative of g at φ_0.

3. EXAMPLES

Optimal control problems have been traditionally described in terms of "control functions". The reader should realize that to every control function corresponds a dynamically feasible mode of operation of the control system. In order to apply the results of Chapter 2 the reader should first make sure that the control problem is stated in terms of a family Γ of modes of operation and not in terms of a family of control functions.

The second preparatory step required for the application of the results of Chapter 2 is to make sure that the given control problem has been written in such a way that the mechanism by which a solution $\varphi \in C(X)$ of some $f \in \Gamma$ is evaluated consists entirely of looking at the value of $g[\varphi]$ where g is some given continuous function from $C(X)$ into some Banach space Z. In classical calculus of variations, for instance, this is done by writing the given problem under its Mayer form.

The scope of applications of the results of Chapter 2 is very large: to the best of my knowledge it includes all the first order conditions in calculus of variations and optimal control theory for problems described by ordinary differential equations when the data is differentiable. In the present chapter I shall only consider seven prototypic problems. Although the description of each of the seven problems will be given independently of the other six, the reader should realize that there is a considerable amount of overlap between those seven problems: Problem 1 is a particular case of Problems 2, 3, 4, 5, 6 and 7; Problems 2 and 6 are closely related; Problems 3 and 7 are closely related; Problems 5, 6 and 7 have almost identical statements; Problems 5 and 7 are dual of each other. The reader should also realize that any combination of features from those seven problems can be used to create a new problem to which the results of Chapter 2 are applicable and that Chapter 2 is in itself the study of the problem combining all those features.

I hope that, after having read those seven examples, the reader will be convinced of the advantage of the formulation of Chapter 2 and that it will not be necessary to return to those specific examples when we shall study problems with nondifferentiable data.

For Problems 1, 2, 3 and 4 one does not need the full strength of the results of Chapter 2 but only the results corresponding to the particular case in which the space Z is a finite dimensional Euclidean space and the measures involved have finite support. In each of the five problems we assume that we are given a family Γ of functions from $X \times [0, 1]$ into X, an element $f_0 \in \Gamma$ and a solution $\varphi_0 \in C(X)$ of f_0 such that Assumptions H1, H2 and H4 hold. As in Chapter 2, I shall denote by A the set of all $\varphi \in C(X)$ such that φ is a solution for some $f \in \Gamma$.

PROBLEM 1 (Standard Optimal Control Problem). We are given nonnegative integers m and μ, and a function $h = \left(h_{-\mu}, \ldots, h_{-1}, h_0, h_1, \ldots, h_m\right)$ from

$X \times X$ into $R^{\mu+m+1}$. We assume that φ_0 is an optimal solution of the following problem: minimize $h_0[\varphi[0], \varphi[1]]$ over the set A subject to the constraints $h_i[\varphi[0], \varphi[1]] \leq 0$ for $i = -\mu, \ldots, -1$ and $h_i[\varphi[0], \varphi[1]] = 0$ for $i = 1, \ldots, m$. In other words if

$$\tilde{A} = \{\varphi : \varphi \in A, h_i[\varphi[0], \varphi[1]] \leq 0 \text{ for } i = -\mu, \ldots, -1 \text{ and }$$
$$h_i[\varphi[0], \varphi[1]] = 0 \text{ for } i = 1, \ldots, m\}$$

we assume that $\varphi_0 \in \tilde{A}$ and that $h_0[\varphi_0[0], \varphi_0[1]] \leq h_0[\varphi[0], \varphi[1]]$ for all $\varphi \in \tilde{A}$. We let $Z = R^{\mu+m+1}$, we define $g = (g_{-\mu}, \ldots, g_m) : C(X) \to R^{\mu+m+1}$ by $g[\varphi] = h[\varphi[0], \varphi[1]]$ and the set

$$K = \{(\alpha_{-\mu}, \ldots, \alpha_m) : \alpha_i \leq 0 \text{ for } i = -\mu, \ldots, -1; \alpha_0 < g_0[\varphi_0];$$
$$\alpha_i = 0 \text{ for } i = 1, \ldots, m\}.$$

We have then $g[\varphi_0] \in \overline{K}$ and $g[A] \cap K = \emptyset$. We assume that h is continuous in a neighborhood of $(\varphi_0[0], \varphi_0[1])$ and admits a derivative (a, b) at $(\varphi_0[0], \varphi_0[1])$. The function g is uniformly continuous in a neighborhood of φ_0 and admits as Fréchet derivative at φ_0 the measure ω consisting entirely of an atom of value a at the point 0 and of an atom of value b at the point 1. By applying Proposition 2.1 to this problem we obtain:

PROPOSITION 3.1. *If φ_0 is an optimal solution for Problem 1 then there exist a nonzero element $q = (q_{-\mu}, \ldots, q_m) \in (R^{\mu+m+1})^*$ and an absolutely continuous function $p : [0, 1] \to X^*$ such that*

(1) $q_i \leq 0$ *for* $i = -\mu, \ldots, 0$,

(2) $q_i h_i[\varphi_0[0], \varphi_0[1]] = 0$ *for* $i = -\mu, \ldots, -1$,

(3) $p[0] = -qa$,

(4) $\dot{p}[t] = -p[t]E[t]$ *for almost all* $t \in [0, 1]$,

(5) $p[1] = qb$,

(6) $\int_0^1 p[t](f_0[\varphi_0[t], t] - f[\varphi_0[t], t])dt \geq 0$ *for all* $f \in \Gamma$.

REMARK. The function p of Proposition 3.1 and the function p_q of Proposition 2.1 are related as follows:

$$p[t] = p_q[t] \text{ for all } t \in (0, 1],$$

$$p[0] = p_q[0+] = \lim_{t \to 0+} p_q[t] .$$

PROBLEM 2 (Finite-Dimensional Vector-Valued Objective: Non Domination). We are given nonnegative integers m, μ and k (with $\mu \geq k$) and a function $h = (h_{-\mu}, \ldots, h_{-k}, \ldots, h_{-1}, h_0, \ldots, h_m)$ from $X \times X$ into $R^{\mu+m+1}$. We assume that φ_0 is an optimal solution of a problem which we shall first describe vaguely as follows: "minimize the vector" $(h_{-k}[\varphi[0], \varphi[1]], \ldots, h_0[\varphi[0], \varphi[1]])$ over the set A subject to the constraints $h_i[\varphi[0], \varphi[1]] \leq 0$ for $i = -\mu, \ldots, -k-1$ and $h_i[\varphi[0], \varphi[1]] = 0$ for $i = 1, \ldots, m$. More precisely if

$$\tilde{A} = \{\varphi : \varphi \in A, h_i[\varphi[0], \varphi[1]] \leq 0 \text{ for } i = -\mu, \ldots, -k-1$$
$$\text{and } h_i[\varphi[0], \varphi[1]] = 0 \text{ for } i = 1, \ldots, m\}$$

we assume that $\varphi_0 \in \tilde{A}$ and that there is no $\varphi \in \tilde{A}$ such that

$$h_i[\varphi[0], \varphi[1]] \leq h_i[\varphi_0[0], \varphi_0[1]] \quad \text{for all } i \in \{-k, \ldots, 0\}$$

and

$$h_i[\varphi[0], \varphi[1]] < h_i[\varphi_0[0], \varphi_0[1]] \quad \text{for some } i \in \{-k, \ldots, 0\} .$$

We let $Z = R^{\mu+m+1}$, we define $g = (g_{-\mu}, \ldots, g_m) : C(X) \to R^{\mu+m+1}$ by $g[\varphi] = h[\varphi[0], \varphi[1]]$ and the set

$$K = \{(\alpha_{-\mu}, \ldots, \alpha_m) : \alpha_i \leq 0 \text{ for } i = -\mu, \ldots, -k-1;$$
$$\alpha_i < g_i[\varphi_0] \text{ for } i = -k, \ldots, 0; \alpha_i = 0 \text{ for } i = 1, \ldots, m\} .$$

We have then $g[\varphi_0] \in \overline{K}$ and $g[A] \cap K = \emptyset$. We assume that h is continuous in a neighborhood of $(\varphi_0[0], \varphi_0[1])$ and admits a derivative (a, b) at $(\varphi_0[0], \varphi_0[1])$. The function g is uniformly continuous in a neighborhood of φ_0 and admits as Fréchet derivative at φ_0 the measure ω consisting entirely of an atom of value a at the point 0 and of an atom of value b at the point 1. By applying Proposition 2.1 to this problem we obtain:

PROPOSITION 3.2. *If φ_0 is an optimal solution for Problem 2 then there exist a nonzero element* $q = (q_{-\mu}, \ldots, q_m) \in (R^{\mu+m+1})*$ *and an absolutely continuous function* $p : [0, 1] \to X*$ *such that*

(1) $q_i \leq 0$ *for* $i = -\mu, \ldots, 0$,

(2) $q_i h_i[\varphi_0[0], \varphi_0[1]] = 0$ *for* $i = -\mu, \ldots, -k-1$,

(3) $p[0] = -qa$,

(4) $\dot{p}[t] = -p[t]E[t]$ *for almost all* $t \in [0, 1]$,

(5) $p[1] = qb$,

(6) $\int_0^1 p[b]\big(f_0[\varphi_0[t], t] - f[\varphi_0[t], t]\big)dt \geq 0$ *for all* $f \in \Gamma$.

PROBLEM 3 (Finite-Dimensional Vector-Valued Objective: Minimax). We are given nonnegative integers m, μ and k (with $\mu \geq k$) and a function $h = (h_{-\mu}, \ldots, h_{-k}, \ldots, h_{-1}, h_0, h_1, \ldots, h_m)$ from $X \times X$ into $R^{\mu+m+1}$. We assume that φ_0 is an optimal solution of the following problem: minimize $\max\{h_{-k}[\varphi[0], \varphi[1]], \ldots, h_0[\varphi[0], \varphi[1]]\}$ over the set A subject to the constraints $h_i[\varphi[0], \varphi[1]] \leq 0$ for $i = -\mu, \ldots, -k-1$ and $h_i[\varphi[0], \varphi[1]] = 0$ for $i = 1, \ldots, m$. More precisely if

$\tilde{A} = \{\varphi : \varphi \in A, h_i[\varphi[0], \varphi[1]] \leq 0$ for $i = -\mu, \ldots, -k-1$

and $h_i[\varphi[0], \varphi[1]] = 0$ for $i = 1, \ldots, m\}$

we assume that $\varphi_0 \in \tilde{A}$ and that for all $\varphi \in \tilde{A}$ we have

$\max\{h_{-k}[\varphi_0[0], \varphi_0[1]], \ldots, h_0[\varphi_0[0], \varphi_0[1]]\}$

$\leq \max\{h_{-k}[\varphi[0], \varphi[1]], \ldots, h_0[\varphi[0], \varphi[1]]\}$.

We let $Z = R^{\mu+m+1}$, we define $g = (g_{-\mu}, \ldots, g_m) : C(X) \to R^{\mu+m+1}$ by $g[\varphi] = h[\varphi[0], \varphi[1]]$ and the set

$K = \{(\alpha_{-\mu}, \ldots, \alpha_m) : \alpha_i \leq 0$ for $i = -\mu, \ldots, -k-1$;

$\alpha_i < \max\{g_{-k}[\varphi_0], \ldots, g_0[\varphi_0]\}$ for $i = -k, \ldots, 0$; $\alpha_i = 0$ for $i = 1, \ldots, m\}$.

We have then $g[\varphi_0] \in \overline{K}$ and $g[A] \cap K = \emptyset$. We assume that h is continuous in a neighborhood of $(\varphi_0[0], \varphi_0[1])$ and admits a derivative (a, b) at $(\varphi_0[0], \varphi_0[1])$. The function g is uniformly continuous in a neighborhood of φ_0 and admits as Fréchet derivative at φ_0 the measure consisting entirely of an atom of value a at the point 0 and of an atom of value b at the point 1. By applying Proposition 2.1 to Problem 3 we obtain:

PROPOSITION 3.3. *If* φ_0 *is an optimal solution for Problem* 3 *then there exist a nonzero element* $q = (q_{-\mu}, \ldots, q_m) \in (R^{\mu+m+1})^*$ *and an absolutely continuous function* $p : [0, 1] \to X^*$ *such that*

(1) $q_i \leq 0$ *for* $i = -\mu, \ldots, 0$,

(2) $q_i = 0$ *if* $i \in \{-\mu, \ldots, -k-1\}$ *and* $h_i[\varphi_0[0], \varphi_0[1]] < 0$

(3) $q_i = 0$ *if* $i \in \{-k, \ldots, 0\}$ *and*

$h_i[\varphi_0[0], \varphi_0[1]] < \max\{h_{-k}[\varphi_0[0], \varphi_0[1]], \ldots, h_0[\varphi_0[0], \varphi_0[1]]\}$,

(4) $p[0] = -qa$

(5) $\dot{p}[t] = -p[t]E[t]$ *for almost all* $t \in [0, 1]$,

(6) $p[1] = qb$,

(7) $\int_0^1 p[t](f_0[\varphi_0[t], t] - f[\varphi_0[t], t])\, dt \geq 0$ *for all* $f \in \Gamma$.

REMARK 1. With the data of Problems 2 or 3 one could want to find an element $\varphi_0 \in \tilde{A}$ such that $h_i[\varphi_0[0], \varphi_0[1]] \leq h_i[\varphi[0], \varphi[1]]$ for all $i \in \{-k, \ldots, 0\}$ and all $\varphi \in A$. The reader should be persuaded that such a problem (when $k > 0$) does not lead to any separation principle since the corresponding set K is not convex.

REMARK 2. On a finite-dimensional vector space endowed with a fixed base we may define the order $\prec$ by the relation $(x'_1, \ldots, x'_k) \prec (x''_1, \ldots, x''_k)$ if

$$x'_i \leq x''_i \text{ for all } i \in \{1, \ldots, k\}$$

and

$$x'_i < x''_i \text{ for some } i \in \{1, \ldots, k\} .$$

With that in mind we may construct all sorts of vector-valued objective criterions which combine the features of Problems 2 and 3. For instance one might want to find an element $\varphi_0 \in \tilde{A}$ such that there is no $\varphi \in \tilde{A}$ with

$(\max\{h_{-2}[\varphi[0], \varphi[1]], h_{-1}[\varphi[0], \varphi[1]]\}, h_0[\varphi[0], \varphi[1]])$

$\prec (\max\{h_{-2}[\varphi_0[0], \varphi_0[1]], h_{-1}[\varphi_0[0], \varphi_0[1]]\}, h_0[\varphi_0[0], \varphi_0[1]])$.

REMARK 3. Problem 3 can also be expressed in the form of Problem 1 by defining a new function $\tilde{h}_0$ from $X \times X$ into R^1 by

$$\tilde{h}_0[x', x''] = \max\{h_{-k}[x', x''], \ldots, h_0[x', x'']\} .$$

Unfortunately the function $\tilde{h}_0$ might fail to be differentiable at $(\varphi_0[0], \varphi_0[1])$ even when the functions $h_{-k}, \ldots, h_0$ are assumed to be differentiable at $(\varphi_0[0], \varphi_0[1])$ and Proposition 3.1 is not applicable to that case. In the section of the present paper devoted to optimization problems with nondifferentiable data I shall give an alternate approach to Problem 3 based on the consideration of the non-differentiable function $\tilde{h}_0$.

PROBLEM 4 (Intermediary Constraints). We are given elements $0 < \tau_1 < \tau_2 < \ldots < \tau_k < 1$, nonnegative integers m and μ, and a function $h = (h_{-\mu}, \ldots, h_0, \ldots, h_m)$ from X^k into $R^{\mu+m+1}$. We assume that φ_0 is an optimal solution of the following problem: minimize $h_0[\varphi[\tau_1], \ldots, \varphi[\tau_k]]$ over the set A subject to the constraints $h_i[\varphi[\tau_1], \ldots, \varphi[\tau_k]] \leq 0$ for $i = -\mu, \ldots, -1$ and $h_i[\varphi[\tau_1], \ldots, \varphi[\tau_k]] = 0$ for $i = 1, \ldots, m$. In other words if

$$\tilde{A} = \{\varphi : \varphi \in A, h_i[\varphi[\tau_1], \ldots, \varphi[\tau_k]] \leq 0 \text{ for } i = -\mu, \ldots, -1$$
$$\text{and } h_i[\varphi[\tau_1], \ldots, \varphi[\tau_k]] = 0 \text{ for } i = 1, \ldots, m\}$$

we assume that $\varphi_0 \in \tilde{A}$ and that $h_0[\varphi_0[\tau_1], \ldots, \varphi_0[\tau_k]] \leq h_0[\varphi[\tau_1], \ldots, \varphi[\tau_k]]$ for all $\varphi \in \tilde{A}$. We let $Z = R^{\mu+m+1}$, we define $g = (g_{-\mu}, \ldots, g_m) : C(X) \to R^{\mu+m+1}$ by $g[\varphi] = h[\varphi[\tau_1], \ldots, \varphi[\tau_k]]$ and the set

$$K = \{(\alpha_{-\mu}, \ldots, \alpha_m) : \alpha_i \leq 0 \text{ for } i = -\mu, \ldots, -1;$$
$$\alpha_0 < g_0[\varphi_0]; \alpha_i = 0 \text{ for } i = 1, \ldots, m\}.$$

We have then $g[\varphi_0] \in \overline{K}$ and $g[A] \cap K = \emptyset$. We assume that h is continuous in a neighborhood of $(\varphi_0[\tau_1], \ldots, \varphi_0[\tau_k])$ and admits a derivative $(a_1, \ldots, a_k)$ at $(\varphi_0[\tau_1], \ldots, \varphi_0[\tau_k])$. The function g is uniformly continuous in a neighborhood of φ_0 and admits as Fréchet derivative at φ_0 a measure consisting entirely of atoms of values a_j at the point τ_j for all $j \in \{1, \ldots, k\}$. By applying Proposition 2.1 to Problem 4 we obtain:

PROPOSITION 3.4. *If φ_0 is an optimal solution for Problem 4 then there exist a nonzero element* $q = (q_{-\mu}, \ldots, q_m) \in (R^{\mu+m+1})^*$ *and a function* $p : [0, 1] \to X^*$ *such that*

(1) $q_i \leq 0$ *for* $i = -\mu, \ldots, 0$,

(2) $q_i h_i[\varphi_0[\tau_1], \ldots, \varphi_0[\tau_k]] = 0$ *for* $i = -\mu, \ldots, -1$,

(3) p *is absolutely continuous on* $[0, 1] \sim \{\tau_1, \ldots, \tau_k\}$, *left continuous on* $(0, 1]$ *and admits a right limit for all* $t \in [0, 1]$,

(4) $p[t] = 0$ *for all* $t \in [0, \tau_1] \cup (\tau_k, 1]$,

(5) $p[\tau_j+] - p[\tau_j] = -qa_j$ *for every* $j \in \{1, \ldots, k\}$,

(6) $\dot{p}[t] = -p[t]E[t]$ *for almost all* $t \in [0, 1]$,

(7) $\int_0^1 p[t](f_0[\varphi_0[t], t] - f[\varphi_0[t], t]) dt \geq 0$ for all $f \in \Gamma$.

REMARK. From relation (5) we see that each functional $h_i[\varphi_0[\tau_1], \ldots, \varphi_0[\tau_k]]$ contributes $-q_i a_j$ to the jump of the adjoint variable p at the point τ_j. An extension of Problem 4 to constraints at an infinite number of intermediary points will be given in Problem 5.

PROBLEM 5 (Bounded Phase Constraints). We are given nonnegative integers m and μ, a function $h = (h_{-\mu}, \ldots, h_{-1}, h_0, h_1, \ldots, h_m)$ from $X \times X$ into $R^{\mu+m+1}$, a closed set $I \subset [0, 1]$, and a function ψ from $X \times I$ into R^1. We assume that φ_0 is an optimal solution of the following problem: minimize $h_0[\varphi[0], \varphi[1]]$ over the set A subject to the constraints $h_i[\varphi[0], \varphi[1]] \leq 0$ for $i = -\mu, \ldots, -1$; $h_i[\varphi[0], \varphi[1]] = 0$ for $i = 1, \ldots, m$ and $\psi[\varphi[t], t] \leq 0$ for all $t \in I$. In other words if

$\tilde{A} = \{\varphi : \varphi \in A, h_i[\varphi[0], \varphi[1]] \leq 0$ for $i = -\mu, \ldots, -1$;

$\quad h_i[\varphi[0], \varphi[1]] = 0$ for $i = 1, \ldots, m$ and $\psi[\varphi[t], t] \leq 0$ for all $t \in I\}$

we assume that $\varphi_0 \in \tilde{A}$ and that $h_0[\varphi_0[0], \varphi_0[1]] \leq h_0[\varphi[0], \varphi[1]]$ for all $\varphi \in \tilde{A}$.

Let $Z_u = R^{\mu+m+1}$, let Z_v denote the Banach space of all real valued continuous functions on I endowed with the sup norm and let $Z = Z_u \times Z_v$. We define $g = (g_{-\mu}, \ldots, g_m, g_{m+1}) : C(X) \to Z$ by the relation $g_i[\varphi] = h_i[\varphi[0], \varphi[1]]$ for $i = -\mu, \ldots, m$ and $g_{m+1}[\varphi][t] = \psi[\varphi[t], t]$ for every $t \in I$. Let

$K_u = \{(\alpha_{-\mu}, \ldots, \alpha_m) : \alpha_i \leq 0$ for $i = -\mu, \ldots, -1$;

$\qquad \alpha_0 < g_0[\varphi_0]; \alpha_i = 0$ for $i = 1, \ldots, m\}$,

let K_v be the set of all elements $\rho \in Z_v$ such that $\rho[t] \leq 0$ for all $t \in I$ and let $K = K_u \times K_v$. Please note that K is substantial since Z_u is finite dimensional and K_v has a nonempty interior. We have then $g[\varphi_0] \in \overline{K}$ and $g[A] \cap K = \emptyset$. We assume that h is continuous in a neighborhood of $(\varphi_0[0], \varphi_0[1])$ and admits a derivative (a, b) at $(\varphi_0[0], \varphi_0[1])$. We assume also that for some $\varepsilon > 0$ the function ψ is continuous on the set

$$S \triangleq \{(x, t) : t \in I, x \in X, |x - \varphi_0[t]| < \varepsilon\}$$

and that there exists a continuous function $D : S \to X^*$ such that for every

$(x, t) \in S$ the element $D[x, t]$ is the derivative of $\psi[x, t]$ with respect to x. The function g is uniformly continuous in a neighborhood of φ_0 and admits as Fréchet derivative at φ_0 the pair (ω, γ) where ω is a measure consisting entirely of an atom of value a at the point 0 and of an atom of value b at the point 1 and where $\gamma \in B(C(X), Z_\nu)$ is given by $(\gamma\varphi)[t] = D[\varphi_0[t], t]\varphi[t]$ for every $\varphi \in C(X)$ and every $t \in I$. By applying Proposition 2.1 to Problem 5 we obtain:

PROPOSITION 3.5. *If φ_0 is an optimal solution for Problem 4 then there exist an element $q = (q_{-\mu}, \ldots, q_m) \in (R^{\mu+m+1})^*$, a Borel measure ν on $[0, 1]$ and a left continuous function of bounded variation $p : [0, 1] \to X^*$ such that*

(1) $(q, \nu) \neq 0$,

(2) $q_i \leq 0$ for $i = -\mu, \ldots, 0$,

(3) $q_i h_i[\varphi_0[0], \varphi_0[1]] = 0$ for $i = -\mu, \ldots, -1$,

(4) *the support of ν is a subset of* $\{t : t \in I, \psi[\varphi_0[t], t] = 0\}$,

(5) $p[0+] = -qa - \nu(\{0\})D[\varphi_0[0], 0]$,

(6) $p[1] = qb + \nu(\{1\})D[\varphi_0[1], 1]$,

(7) $p[t_1] - p[t_0] = -\int_{[t_0, t_1)} (p[t]E[t]dt + \nu[dt]D[\varphi_0[t], t])$ for all $t_0 < t_1$ in $[0, 1]$,

(8) $\int_0^1 p[t](f_0[\varphi_0[t], t] - f[\varphi_0[t], t])dt \geq 0$ for all $f \in \Gamma$.

PROBLEM 6 (Infinite-Dimensional Vector-Valued Objective: Non Domination). We are given nonnegative integers m and μ, a function $h = (h_{-\mu}, \ldots, h_{-1}, h_0, h_1, \ldots, h_m)$ from $X \times X$ into $R^{\mu+m+1}$, a closed set $I \subset [0, 1]$ and a function ψ from $X \times I$ into R^1. We assume that φ_0 is an optimal solution of a problem which we shall first describe vaguely as follows: obtain the smallest function $t \to \psi[\varphi[t], t]$ over the set of all $\varphi \in A$ subject to the constratins $h_i[\varphi[0], \varphi[1]] \leq 0$ for $i = -\mu, \ldots, 0$ and $h_i[\varphi[0], \varphi[1]] = 0$ for $i = 1, \ldots, m$. More precisely if

$\tilde{A} = \{\varphi : \varphi \in A, h_i[\varphi[0], \varphi[1]] \leq 0$ for $i = -\mu, \ldots, 0;$

$h_i[\varphi[0], \varphi[1]] = 0$ for $i = 1, \ldots, m\}$

we assume that $\varphi_0 \in \tilde{A}$ and that there is no $\varphi \in \tilde{A}$ such that

$$\psi[\varphi[t], t] \le \psi[\varphi_0[t], t] \quad \text{for all} \quad t \in I$$

and

$$\psi[\varphi[t], t] < \psi[\varphi_0[t], t] \quad \text{for some} \quad t \in I .$$

Let $Z_u = R^{\mu+m+1}$, let Z_v denote the Banach space of all real valued continuous functions on I endowed with the sup norm and let $Z = Z_u \times Z_v$. We define $g = (g_{-\mu}, \ldots, g_m, g_{m+1}) : C(X) \to Z$ by the relation $g_i[\varphi] = h_i[\varphi[0], \varphi[1]]$ for $i = -\mu, \ldots, m$ and $g_{m+1}[\varphi][t] = \psi[\varphi[t], t]$ for every $t \in I$. Let $K_u = \{(\alpha_{-\mu}, \ldots, \alpha_m) : \alpha_i \le 0 \text{ for } i = -\mu, \ldots, 0; \alpha_i = 0 \text{ for } i = 1, \ldots, m\}$, let K_v be the set of all elements $\rho \in Z_v$ such that $\rho[t] < \psi[\varphi_0[t], t]$ for all $t \in I$ and let $K = K_u \times K_v$. Please note that K is substantial since Z_u is finite dimensional and K_v has a nonempty interior. We have then $g[\varphi_0] \in \overline{K}$ and $g[A] \cap K = \emptyset$. We assume that h is continuous in a neighborhood of $(\varphi_0[0], \varphi_0[1])$ and admits a derivative (a, b) at $(\varphi_0[0], \varphi_0[1])$. We assume that for some $\varepsilon > 0$ the function ψ is continuous on the set $S \triangleq \{(x, t) : t \in I, x \in X, |x-\varphi_0[t]| < \varepsilon\}$ and that there exists a continuous function $D : S \to X^*$ such that for every $(x, t) \in S$ the element $D[x, t]$ is the derivative of $\psi[x, t]$ with respect to x. The function g is uniformly continuous in a neighborhood of φ_0 and admits as Fréchet derivative at φ_0 the pair (ω, γ) where ω is a measure consisting entirely of an atom of value a at the point 0 and of an atom of value b at the point 1 and where $\gamma \in B(C(X), Z_v)$ is given by $(\gamma\varphi)[t] = D[\varphi_0[t], t]\varphi[t]$ for every $\varphi \in C(X)$ and every $t \in I$. By applying Proposition 2.1 to Problem 6 we obtain:

PROPOSITION 3.6. *If φ_0 is an optimal solution for Problem 6 then there exist an element $q = (q_{-\mu}, \ldots, q_m) \in (R^{\mu+m+1})^*$, a nonnegative Borel measure ν on $[0, 1]$ and a left continuous function of bounded variation $p : [0, 1] \to X^*$ such that*

(1) $(q, \nu) \ne 0$,

(2) $q_i \le 0$ for $i = -\mu, \ldots, 0$,

(3) $q_i h_i[\varphi_0[0], \varphi_0[1]] = 0$ for $i = -\mu, \ldots, -1, 0$,

(4) *the support of ν is a subset of I,*

(5) $p[0+] = -qa - \nu(\{0\})D[\varphi_0[0], 0]$,

(6) $p[1] = qb + \nu(\{1\})D[\varphi_0[1], 1]$,

(7) $p[t_1] - p[t_0] = - \int_{[t_0,t_1]} (p[t]E[t]dt + \nu[dt]D[\varphi_0[t], t])$ for all

$t_0 < t_1$ in $[0, 1]$,

(8) $\int_0^1 p[t](f_0[\varphi_0[t], t] - f[\varphi_0[t], t])dt \geq 0$ for all $f \in \Gamma$.

PROBLEM 7 (Infinite-Dimensional Vector-Valued Objective: Minimax). We are given nonnegative integers m and μ , a function $h = (h_{-\mu}, \ldots, h_{-1}, h_0, h_1, \ldots, h_m)$ from $X \times X$ into $R^{\mu+m+1}$, a closed set $I \subset [0, 1]$ and a function ψ from $X \times I$ into R^1. We assume that φ_0 is an optimal solution of the following problem: minimize $\sup_{t \in I} \psi[\varphi[t], t]$ over the set A subject to the constraints $h_i[\varphi[0], \varphi[1]] \leq 0$ for $i = -\mu, \ldots, 0$ and $h_i[\varphi[0], \varphi[1]] = 0$ for $i = 1, \ldots, m$. In other words if

$\tilde{A} = \{\varphi : \varphi \in A, h_i[\varphi[0], \varphi[1]] \leq 0$ for $i = -\mu, \ldots, 0;$

$h_i[\varphi[0], \varphi[1]] = 0$ for $i = 1, \ldots, m\}$

we assume that $\varphi_0 \in \tilde{A}$ and that $\sup_{t \in I} \psi[\varphi_0[t], t] \leq \sup_{t \in I} \psi[\varphi[t], t]$ for all $\varphi \in \tilde{A}$.

Let $Z_u = R^{\mu+m+1}$, let Z_v denote the Banach space of all real valued continuous functions on I endowed with the sup norm and let $Z = Z_u \times Z_v$. We define $g = (g_{-\mu}, \ldots, g_m, g_{m+1}) : C(X) \to Z$ by the relation $g_i[\varphi] = h_i[\varphi[0], \varphi[1]]$ for $i = -\mu, \ldots, m$ and $g_{m+1}[\varphi][t] = \psi[\varphi[t], t]$ for every $t \in I$. Let

$K_u = \{(\alpha_{-\mu}, \ldots, \alpha_m) : \alpha_i \leq 0$ for $i = -\mu, \ldots, 0; \alpha_i = 0$ for $i = 1, \ldots, m\}$,

let K_v be the set of all elements $\rho \in Z_v$ such that $\rho[t] < \sup_{\tau \in I} \psi[\varphi_0[\tau], \tau]$ for all $t \in I$ and let $K = K_u \times K_v$. Please note that the set K is substantial since Z_u is finite dimensional and K_v has a nonempty interior. We have then $g[\varphi_0] \in \overline{K}$ and $g[A] \cap K = \emptyset$. We assume that h is continuous in a neighborhood of $(\varphi_0[0], \varphi_0[1])$ and admits a derivative (a, b) at $(\varphi_0[0], \varphi_0[1])$. We assume also that for some $\varepsilon > 0$ the function ψ is continuous on the set

$S \triangleq \{(x, t) : t \in I, x \in X, |x - \varphi_0[t]| < \varepsilon\}$

and that there exists a continuous function $D : S \to X^*$ such that for every $(x, t) \in S$ the element $D[x, t]$ is the derivative of $\psi[x, t]$ with respect to x. The function g is uniformly continuous in a neighborhood of φ_0 and admits as Fréchet derivative at φ_0 the pair (ω, γ) where ω is a measure consisting entirely of an atom of value a at the point 0 and of an atom of value b at the point 1 and where $\gamma \in B(C(X), Z_\nu)$ is given by $(\gamma\varphi)[t] = D[\varphi_0[t], t]\varphi[t]$ for every $\varphi \in C(X)$ and every $t \in I$. By applying Proposition 2.1 to Problem 7 we obtain:

PROPOSITION 3.7. *If φ_0 is an optimal solution for Problem 7 then there exist an element $q = (q_{-\mu}, \ldots, q_m) \in (R^{\mu+m+1})^*$, a nonnegative Borel measure ν on $[0, 1]$ and a left continuous function of bounded variation $p : [0, 1] \to X^*$ such that*

(1) $(q, \nu) \neq 0$,

(2) $q_i \leq 0$ for $i = -\mu, \ldots, 0$,

(3) $q_i h_i [\varphi_0[0], \varphi_0[1]] = 0$ for $i = -\mu, \ldots, -1, 0$,

(4) *the support of ν is a subset of*
$$\{\tau : \tau \in I, \psi[\varphi_0[\tau], \tau] = \sup_{t \in I} \psi[\varphi_0[t], t]\},$$

(5) $p[0+] = -qa - \nu(\{0\})D[\varphi_0[0], 0]$,

(6) $p[1] = qb + \nu(\{1\})D[\varphi_0[1], 1]$,

(7) $p[t_1] - p[t_0] = -\int_{[t_0, t_1]} (p[t]E[t]dt + \nu[dt]D[\varphi_0[t], t])$ for all $t_0 < t_1$ in $[0, 1]$,

(8) $\int_0^1 p[t](f_0[\varphi_0[t], t] - f[\varphi_0[t], t])dt \geq 0$ for all $f \in \Gamma$.

REMARK. The three remarks given after Problem 3 apply also to Problems 6 and 7.

4. Weak* Compact Normal Set for Substantial Convex Sets

Let Y be a Banach space and let Y^* be the space of all continuous linear functionals on Y.

If Y is finite-dimensional, K is a convex subset of Y, $y_0 \in Y$, then there exists a compact set $Q \subset Y^*$ such that $0 \notin Q$ and such that for every $q_1 \in Y^*$

with $q_1 \neq 0$ and $q_1(y-y_0) \geq 0$ for all $y \in K$ there exist some $q_2 \in Q$ and some $t > 0$ with $q_1 = tq_2$. The set $Q = \{q : q \in Y^*, |q| = 1\}$ satisfies all the specified requirements. Please note that Q does not depend on K or y_0.

In Proposition 4.1 of the present chapter, I will prove a similar result for the case in which the space Y is not necessarily finite dimensional. The set K will be required to be substantial (the definition given in Chapter 1 will be recalled below) and the set Q will then be a weak* compact set depending on K and y_0.

DEFINITION (Chapter 1). A convex subset K of a Banach space Y is said to be *substantial* if K has a nonempty relative interior with respect to a closed affine subspace of Y of finite codimension, that is, if there exist an element $y^* \in K$, a real number $\varepsilon > 0$, an integer $m \geq 0$ and a continuous affine function $p : Y \to R^m$ such that

(1) $K \subset \{y : y \in Y, p[y] = 0\}$,

(2) $\{y : y \in Y, p[y] = 0, |y-y^*| < \varepsilon\} \subset K$.

REMARK. A nonempty convex subset of a finite dimensional Euclidean space is always substantial.

PROPOSITION 4.1. *If Y is a Banach space, $K \subset Y$ is convex and substantial, $y_0 \in Y$ then there exists a weak* compact set $Q \subset Y^*$ such that $0 \notin Q$ and such that for every $q_1 \in Y^*$ with $q_1 \neq 0$ and $q_1(y-y_0) \geq 0$ for all $y \in K$ there exist some $q_2 \in Q$ and some $t > 0$ with $q_1 = tq_2$.*

Proof of Proposition 4.1. We may assume without loss of generality that $y_0 \in K$; if necessary replace K by $co(K \cup \{y_0\})$. We may also assume that $0 \in K$; if necessary perform an appropriate translation of the data. Since the set K is substantial we may assume that

(1) $Y = Y_- \times Y_+$ where Y_- is a Banach space and Y_+ is a finite-dimensional Euclidean space,

(2) $K = K_- \times \{0\}$ where K_- is a subset of Y_- with nonempty interior.

We have then $y_0 = (y_0^-, 0)$ for some $y_0^- \in K_-$. Let $\varepsilon > 0$ and let $y_*^- \in K_-$ such that $y^- \in K_-$ whenever $|y^- - y_*^-| < \varepsilon$. Let Q_- be the set of all $q_- \in (Y_-)^*$ such that $q_-(y^- - y_0^-) \geq 0$ for all $y^- \in K_-$ and $q_-(y_*^- - y_0^-) = +1$. The set Q_- is weak* compact since it is weak* closed and bounded in the uniform topology by $1/\varepsilon$. If $\dim(Y_+) > 0$ let Q_+ be the set of all $q_+ \in (Y_+)^*$ such that $|q_+| = +1$. The set Q_+ is compact. Let $Q \subset Y^*$ be defined as follows:

A. if $\dim(Y_+) = 0$ let $Q = Q_-$,

B. if $\dim(Y_+) > 0$ and $Q_- = \emptyset$ let $Q = \{(0, q_+) : q_+ \in Q_+\}$,

C. if $\dim(Y_+) > 0$ and $Q_- \neq \emptyset$ let

$$Q = \{(\mu q_-, (1-\mu)q_+) : \mu \in [0, 1], q_- \in Q_-, q_+ \in Q_+\} .$$

The set Q has all the required properties. This concludes the proof of Proposition 4.1.

5. Chattering Principle

Let X be a finite dimensional Euclidean space. I have already defined $C(X)$ as the Banach space of all continuous functions from $[0, 1]$ into X with the norm $|\varphi| = \sup\limits_{t \in [0,1]} |\varphi[t]|$.

I assume that for some $\varphi_0 \in C(X)$ and some $f_0 : X \times [0, 1] \to X$ we have

$$\varphi_0[t] = \varphi_0[0] + \int_0^t f_0[\varphi_0[\tau], \tau] d\tau \quad \text{for all } t \in [0, 1] .$$

In the present chapter, I want to study the solution of the integral equation

$$\varphi[t] = \varphi_0[0] + x + \int_0^t f[\varphi[\tau], \tau] d\tau$$

for all $t \in [0, 1]$ when $x \in X$ is "small" and when $f : X \times [0, 1] \to X$ is "close" to f_0 . This chapter is devoted to existence, continuity and convergence properties of solutions. No question of differentiability in either a classical or a "set-valued" sense is touched upon in the present chapter. Proposition 5.1, the Chattering Principle, states that "convexity-under-switching" is almost as good as convexity, that is, that by switching "sufficiently frequently" between modes of operation of a control system one obtains a response very "close" to the response which would be obtained by using some average of those modes of operation. In other words, switches are as good as rheostats.

We are given a number $\rho > 0$, an integrable function $\sigma : [0, 1] \to [0, \infty)$, functions $f_0, f_1, \ldots, f_l : X \times [0, 1] \to X$ which we assume to be of class $\Xi(\rho, \sigma)$ around φ_0 ; see the definition in Chapter 2. Let Y be an l-dimensional space with norm $|(y_1, \ldots, y_l)| = \sum\limits_{i=1,\ldots,l} |y_i|$ and let

$$P = \{y = (y_1, \ldots, y_l) : y \in Y, y_i \geq 0 \text{ for } i = 1, \ldots, l, |y| \leq 1\} .$$

For any $y \in P$ let $f_y : X \times [0, 1] \to X$ be the y affine combination of the functions $f_0, f_1, \ldots, f_l$; see the definition in Chapter 2. For any $y \in P$ and any $j \in \{1, 2, \ldots\}$ let $f_{y,j} : X \times [0, 1] \to X$ be the y, j switching combination of the functions $f_0, f_1, \ldots, f_l$; see the definition in Chapter 2. We remark that the functions f_y and $f_{y,j}$ are of class $\Xi(\rho, \sigma)$ around φ_0 whenever $y \in P$ and $j \in \{1, 2, \ldots\}$. We shall use the notation $|\sigma| = \int_0^1 \sigma[t]dt$. Let $\alpha \in (0, 1]$ be such that $\alpha e^{|\sigma|}(1+2|\sigma|) < \rho$ and $e^{|\sigma|}\left(\alpha+2\int_T \sigma[t]dt\right) < \rho$ for all measurable sets $T \subset [0, 1]$ with $\mu(T) \leq \alpha$. Let $\Delta = \{(x, y) : x \in X, |x| \leq \alpha, y \in P, |y| \leq \alpha\}$.

An elementary application[14] of Picard's Method and Gronwall's inequality allows me to assert that

(1) for every $(x, y) \in \Delta$ there exists a unique element $\varphi_{x,y} \in N(\varphi_0, \rho)$ such that $\varphi_{x,y}[t] = \varphi_0[0] + x + \int_0^t f_y[\varphi_{x,y}[\tau], \tau]d\tau$ for every $t \in [0, 1]$; moreover, the mapping $(x, y) \to \varphi_{x,y}$ is continuous;

(2) for every $(x, y) \in \Delta$ and every $j \in \{1, 2, \ldots\}$ there exists a unique element $\varphi_{x,y,j} \in N(\varphi_0, \rho)$ such that

$$\varphi_{x,y,j}[t] = \varphi_0[0] + x + \int_0^t f_{y,j}[\varphi_{x,y,j}[\tau], \tau]d\tau$$

for every $t \in [0, 1]$; moreover, the mappings $(x, y) \to \varphi_{x,y,j}$, $j = 1, 2, \ldots$ are uniformly equicontinuous on Δ.

PROPOSITION 5.1 (Chattering Principle). *The mappings* $(x, y) \to \varphi_{x,y,j}$, $j = 1, 2, \ldots$, *converge uniformly on the set* Δ *to the mapping* $(x, y) \to \varphi_{x,y}$ *as* j *tends to* $+\infty$.

LEMMA 5.1. *If* $\bar{g}_0, \bar{g}_1, \ldots, \bar{g}_l$ *are step functions from* $[0, 1]$ *into* X, *if for all* $y \in P$ *we denote by* $\bar{g}_y$ *the* y *affine combination of the functions* $\bar{g}_0, \bar{g}_1, \ldots, \bar{g}_l$, *if for all* $y \in P$ *and* $j \in \{1, 2, \ldots\}$ *we denote by* $\bar{g}_{y,j}$ *the* y, j *switching combination of the functions* $\bar{g}_0, \bar{g}_1, \ldots, \bar{g}_l$ *then*

$$\lim_{j \to \infty} \sup_{t \in [0,1], y \in P} \left| \int_0^t (\bar{g}_{y,j}[\tau] - \bar{g}_y[\tau])d\tau \right| = 0.$$

Proof of Lemma 5.1. Given $\varepsilon > 0$ I shall show that for some $j^* \in \{1, 2, \ldots\}$ we have $\left|\int_0^t (\bar{g}_{y,j}[\tau] - \bar{g}_y[\tau])d\tau\right| \leq \varepsilon$ for all $t \in [0, 1]$, $y \in P$ and $j \geq j^*$. Let

---
[14] See Appendix.

$M = \sup\limits_{\tau \in [0,1], i \in \{0,1,\ldots,l\}} |\bar{g}_i[\tau]|$ and let Q be the number of elements in $[0, 1]$ where one at least of the functions $\bar{g}_i$ is discontinuous. Let $j^* \in \{1, 2, \ldots\}$ be such that $2MQ/j^* \leq \varepsilon$. Let $j \geq j^*$. Let A_j be the set of all $k \in \{1, 2, \ldots, j\}$ such that g_i is constant on $\left[\frac{k-1}{j}, \frac{k}{j}\right]$ for all $i \in \{0, 1, \ldots, l\}$ and let $B_j = \{1, 2, \ldots, j\} \sim A_j$. The set B_j contains at most Q elements. For every $k \in A_j$ and every $y \in P$ we have $\left|\int_{(k-1)/j}^{k/j} (\bar{g}_{y,j}[\tau] - \bar{g}_y[\tau]) d\tau\right| = 0$. For every $k \in B_j$ and every $y \in P$ we have $\int_{(k-1)/j}^{k/j} |\bar{g}_{y,j}[\tau] - \bar{g}_y[\tau]| d\tau \leq \frac{2M}{j}$. We have then

$$\left|\int_0^t (\bar{g}_{y,j}[\tau] - \bar{g}_y[\tau]) d\tau\right| \leq \frac{2MQ}{j} \leq \varepsilon$$

for all $t \in [0, 1]$, $y \in P$ and $j \geq j^*$. This concludes the proof of Lemma 5.1.

LEMMA 5.2. *If $g_0, g_1, \ldots, g_l$ are integrable functions from $[0, 1]$ into X, if for all $y \in P$ we denote by g_y the y affine combination of the functions $g_0, g_1, \ldots, g_l$, if for all $y \in P$ and $j \in \{1, 2, \ldots\}$ we denote by $g_{y,j}$ the y, j switching combination of the functions $g_0, g_1, \ldots, g_l$, then*

$$\lim_{j \to \infty} \sup_{t \in [0,1], y \in P} \left|\int_0^t (g_{y,j}[\tau] - g_y[\tau]) d\tau\right| = 0.$$

Proof of Lemma 5.2. Given $\varepsilon > 0$ let $\bar{g}_i$, $i \in \{0, 1, \ldots, l\}$ be step functions such that $\int_0^1 |\bar{g}_i[\tau] - g_i[\tau]| d\tau \leq \varepsilon/3(l+1)$. Let $\bar{g}_y$ be the y affine combination of the $\bar{g}_i$, $i \in \{0, 1, \ldots, l\}$ and let $\bar{g}_{y,j}$ be the y, j switching combination of the $\bar{g}_i$, $i \in \{0, 1, \ldots, l\}$. For every $j \in \{1, 2, \ldots\}$ we have

$\int_0^1 |\bar{g}_y[\tau] - g_y[\tau]| d\tau$ and $\int_0^1 |\bar{g}_{y,j}[\tau] - g_{y,j}[\tau]| d\tau \leq \varepsilon/3$. From Lemma 5.1 we know that there exists a $j^* \in \{1, 2, \ldots\}$ such that for all $j \geq j^*$ we have

$$\left|\int_0^t (\bar{g}_{y,j}[\tau] - \bar{g}_y[\tau]) d\tau\right| \leq \varepsilon/3$$

for all $t \in [0, 1]$ and $y \in P$. For all $j \geq j^*$, $t \in [0, 1]$ and $y \in P$ we have then

$$\left|\int_0^t (g_{y,j}[\tau] - g_y[\tau]) d\tau\right| \leq \left|\int_0^t (\bar{g}_{y,j}[\tau] - \bar{g}_y[\tau]) d\tau\right| + \int_0^1 |\bar{g}_y[\tau] - g_y[\tau]| d\tau$$

$$+ \int_0^1 |\bar{g}_{y,j}[\tau] - g_{y,j}[\tau]| d\tau \leq \varepsilon/3 + \varepsilon/3 + \varepsilon/3 = \varepsilon.$$

This concludes the proof of Lemma 5.2.

Proof of Proposition 5.1. Since Δ is compact and since the mappings $(x, y) \to \varphi_{x,y,j}$, $j = 1, 2, \ldots$ are uniformly equicontinuous on Δ, we know, from Arzela-Ascoli's Theorem, that it is sufficient to prove that for every $(x, y) \in \Delta$ we have $\lim_{j \to \infty} \varphi_{x,y,j} = \varphi_{x,y}$. Indeed, for every $(x, y) \in \Delta$, every $j \in \{1, 2, \ldots\}$ and every $t \in [0, 1]$ we have

$$|\varphi_{x,y,j}[t] - \varphi_{x,y}[t]| \leq \int_0^t \sigma[\tau] |\varphi_{x,y,j}[\tau] - \varphi_{x,y}[\tau]| d\tau + V_{t,j}$$

where

$$V_{t,j} = \left| \int_0^t \left(f_{y,j}[\varphi_{x,y}[\tau], \tau] - f_y[\varphi_{x,y}[\tau], \tau] \right) d\tau \right| .$$

From Gronwall's inequality we then obtain $|\varphi_{x,y,j} - \varphi_{x,y}| \leq e^{|\sigma|} \sup_{t \in [0,1]} V_{t,j}$. Since we know, Lemma 5.2, that $\lim_{j \to \infty} \sup_{t \in [0,1]} V_{t,j} = 0$, it then follows that $\lim_{j \to \infty} |\varphi_{x,y,j} - \varphi_{x,y}| = 0$. This concludes the proof of Proposition 5.1.

6. Dual Integral Equations with Respect to Borel Measures

Let X be a finite dimensional Euclidean space. I have already introduced the notation $B(X, X)$ for the space of all linear mappings from X into X. I shall now introduce some further notations: $BM(X)$ for the space of all bounded measurable functions from $[0, 1]$ into X and $MB(X)$ for the space of all X valued Borel measures on $[0, 1]$.

Let E be a given integrable function from $[0, 1]$ into $B(X, X)$. For every $\psi \in BM(X)$ there exists a unique element $\varphi \in BM(X)$ such that

$$\varphi[t] = \int_0^t E[\tau]\varphi[\tau]d\tau + \psi[t] \quad \text{for all } t \in [0, 1] .$$

This can be seen most easily by remarking that the equivalent integral equation

$$\varphi[t] - \psi[t] = \int_0^t E[\tau](\varphi[\tau] - \psi[\tau])d\tau + \int_0^t E[\tau]\psi[\tau]d\tau \quad \text{for all } t \in [0, 1]$$

admits a unique absolutely continuous solution $\varphi - \psi$.

In Proposition 6.1, I prove some duality results for solutions of adjoint integral equations when the functions ψ are right and left distribution functions of Borel measures.

PROPOSITION 6.1. *For every* $\lambda \in MB(X)$ *the integral equation*

(1) $$q[t] = \int_0^t E[\tau]q[\tau]d\tau + \lambda([0, t]) \quad \text{for all} \quad t \in [0, 1]$$

admits a unique solution $q \in BM(X)$. *For every* $\nu \in MB(X^*)$ *the integral equation*

(2) $$p[\tau] = \int_\tau^1 p[t]E[t]dt + \nu([\tau, 1]) \quad \text{for all} \quad \tau \in [0, 1]$$

admits a unique solution $p \in BM(X^*)$. *Moreover*

(3) $$\int_{[0,1]} \nu[dt]q[t] = \int_{[0,1]} p[\tau]\lambda[d\tau].$$

Proof of Proposition 6.1. The existence and uniqueness of p and q have been already established. It remains to prove relation (3). Although relation (3) can be derived from the formula of integration by parts of a right continuous function of bounded variation with respect to a left continuous function of bounded variation, I prefer to give below a direct proof in terms of Borel measures since I want to avoid all the notational difficulties associated with right and left limits. The formula of integration by parts of functions of bounded variation contains valuation and/or jump terms at 0 and 1 which are automatically incorporated in relation (3).

Let $S = \{(t, \tau) : t \in [0, 1], \tau \in [0, t]\}$ and let $A = \iint_S \nu[dt]\lambda[d\tau]$. We have

$$\int_{[0,1]} \nu[dt]q[t] = \int_{[0,1]} \nu[dt]\left(\int_{[0,t]} (E[\tau]q[\tau]d\tau + \lambda[d\tau])\right)$$

$$= A + \iint_S \nu[dt]E[\tau]q[\tau]d\tau$$

$$= A + \int_{[0,1]} \nu([\tau, 1])E[\tau]q[\tau]d\tau$$

$$= A + \int_{[0,1]} \left(p[\tau] - \int_{[\tau,1]} p[t]E[t]dt\right)E[\tau]q[\tau]d\tau$$

$$= A + \int_{[0,1]} p[\tau]E[\tau]q[\tau]d\tau - \iint_S p[t]E[t]E[\tau]q[\tau]dtd\tau.$$

Similarly, we have

$$\int_{[0,1]} p[\tau]\lambda[d\tau] = \int_{[0,1]} \left(\int_{[\tau,1]} (p[t]E[t]dt + \nu[dt]) \right) \lambda[d\tau]$$

$$= A + \iint_S p[t]E[t]\lambda[d\tau]dt$$

$$= A + \int_{[0,1]} p[t]E[t]\lambda([0, t])dt$$

$$= A + \int_{[0,1]} p[t]E[t]\left[q[t] - \int_{[0,t]} E[\tau]q[\tau]d\tau \right] dt$$

$$= A + \int_{[0,1]} p[t]E[t]q[t]dt - \iint_S p[t]E[t]E[\tau]q[\tau]dtd\tau \ .$$

This concludes the proof of Proposition 6.1.

REMARK. A longer but less mysterious proof of Proposition 6.1 could be given according to the following outline. There exists a unique continuous function $G : S \to B(X, X)$ such that for every $(t, \tau) \in S$ we have

(4) $$G[t, \tau] = I + \int_\tau^t E[\theta]G[\theta, \tau]d\theta \ .$$

For every $(t, \tau) \in S$ we have also

(5) $$G[t, \tau] = I + \int_\tau^t G[t, \theta]E[\theta]d\theta \ .$$

The function $q : [0, 1] \to X$ defined by

(6) $$q[t] = \int_{[0,t]} G[t, \tau]\lambda[d\tau] \quad \text{for every } t \in [0, 1]$$

is a solution of the integral equation (1). Similarly the function $p : [0, 1] \to X^*$ defined by

(7) $$p[\tau] = \int_{[\tau,1]} \nu[dt]G[t, \tau] \quad \text{for every } \tau \in [0, 1]$$

is a solution of the integral equation (2). We have then

(8) $$\int_{[0,1]} \nu[dt]q[t] = \iint_S \nu[dt]G[t, \tau]\lambda[d\tau]$$

and similarly

(9) $$\int_{[0,1]} p[\tau]\lambda[d\tau] = \iint_S \nu[dt]G[t, \tau]\lambda[d\tau] \ .$$

Finally let us note that q is not only bounded integrable but is also a function of bounded variation which is continuous on the right. Similarly p is a function of bounded variation which is continuous on the left.

APPLICATION. In Chapter 2, I apply Proposition 6.1 to X valued measures $\lambda_{x,y}$

defined by $\lambda_{x,y}([0, t]) = x + \left(\int_0^t D[\tau]d\tau\right) y$ for all $t \in [0, 1]$. Note that the measure $\lambda_{x,y}$ is made up of an atom of value x located at $t = 0$ and of an absolutely continuous part with density $D[t]y$. For all $x \in X$, $y \in Y$, I denote by $\bar{\varphi}_{x,y}$ the unique element of $BM(X)$ such that

$$\bar{\varphi}_{x,y}[t] = \int_0^t E[\tau]\bar{\varphi}_{x,y}[\tau]d\tau + x + \left(\int_0^t D[\tau]d\tau\right) y$$

for all $t \in [0, 1]$. It is easy to see that $\bar{\varphi}_{x,y} \in C(X)$ and that the mapping $(x, y) \to \bar{\varphi}_{x,y}$ is linear and continuous, that is, that for some element $H \in B(X \times Y, C(X))$ we have $\bar{\varphi}_{x,y} = H(x, y)$ for all $(x, y) \in X \times Y$. We have then

$$\bigl(H(x, y)\bigr)[t] = \int_0^t E[\tau]\bigl(H(x, y)\bigr)[\tau]d\tau + \lambda_{x,y}([0, t])$$

for all $x \in X$, $y \in Y$ and $t \in [0, 1]$.

In Chapter 2, I consider an X^* valued measure $q\omega$ and I denote by p_q the element of $BM(X^*)$ defined by

$$p_q[\tau] = \int_\tau^1 p_q[t]E[t]dt + q\omega([\tau, 1])$$

for all $\tau \in [0, 1]$. From relation (3) of Proposition 6.1, I thus obtain

$$\int_0^1 q\omega[dt]\bigl(H(x, y)\bigr)[t] = \int_0^1 p_q[t]\lambda_{x,y}[dt]$$

and hence

$$q\omega H(x, y) = p_q[0]x + \left(\int_0^1 p_q[t]D[t]\right) y .$$

Please note that the identity

$$q\omega H(x, y) = \int_0^1 q\omega[dt]\bigl(H(x, y)\bigr)[t]$$

comes from the fact that we identify an element $q\omega \in \bigl(C(X)\bigr)^*$ with the X^* valued measure $q\omega$ representing it according to Riesz Theorem.

7. A Preview of Nondifferentiable Optimization Theory

In Proposition 7.1 of the present chapter I will state a separation principle for a control problem in which the data is not assumed to be differentiable. The proof of Proposition 7.1 is given in Halkin [17]. The data, assumptions and notations for

Proposition 7.1 are the same as in Proposition 2.1 with the exception of Assumptions H3 and H4 which will take the weaker forms H3* and H4* given below.

I will conclude the present chapter by giving in Proposition 7.2 a Carathéodory-John Multiplier Rule for a nondifferentiable mathematical programming problem over a Banach space in the presence of a vector-valued objective function (of possibly infinite dimension) a finite number of equality constratins and a possibly infinite number of inequality constraints.

If X and Y are Banach spaces I recall that $B(X, Y)$ denotes the set of all continuous linear functions from X into Y. On the space $B(X, Y)$, I shall consider the strong operator topology obtained by taking as base all sets of the form $\{\bar{a} : \bar{a} \in B(X, Y), |ax - \bar{a}x| < \varepsilon \text{ for all } x \in S\}$ where $a \in B(X, Y)$, S is a finite subset of X and $\varepsilon > 0$. When Y is the real line R^1, the strong operator topology on $B(X, R^1) = X^*$ is usually called the weak* topology of X^* or the X topology of X^*.

DEFINITION. If X and Y are Banach spaces, if $f : X \to Y$ then a subset A of $B(X, Y)$ is a P-shield for f at $x_0 \in X$ if A is convex and strong operator compact and if for any compact set $\Omega \subset X$ and $\varepsilon > 0$ there exists a $\delta > 0$ such that for any $x \in \Omega \cap N(x_0, \delta)$ there is an $a \in A$ with

$$|f[x_0] + a(x - x_0) - f[x]| \leq \varepsilon |x - x_0| .$$

DEFINITION. If X and Y are Banach spaces, if $f : X \to Y$ then a subset A of $B(X, Y)$ is a shield for f at $x_0 \in X$ if $A = \bigcap_{i=1,2,\ldots} A_i$ where the sets $A_1 \supset A_2 \supset \ldots$ are P-shields for f at x_0.

We see immediately that if $f : X \to Y$ is Fréchet differentiable at $x_0 \in X$ then $\{f'[x_0]\}$ is a shield for f at x_0.

ASSUMPTIONS. H3* The function g is uniformly continuous in a neighborhood of φ_0 and admits a shield $\Omega \subset B(C(X), Z)$ at φ_0.

H4* For every $t \in [0, 1]$ a set $\Phi[t] \subset B(X, X)$ is given such that

(a) for every $t \in [0, 1]$ the set $\Phi[t]$ is a shield for the mapping $x \to f_0[x, t]$ at the point $x = \varphi_0[t]$;

(b) for some integrable function $\sigma : [0, 1] \to [0, \infty)$ we have $|\Phi[t]| \leq \sigma[t]$ for all $t \in [0, 1]$;

(c) the mapping $t \to \Phi[t]$ is measurable, that is, for every open set $S \subset B(X, X)$ the set $\{t : t \in [0, 1], \Phi[t] \cap S \neq \emptyset\}$ is measurable.

PROPOSITION 7.1. *If* $g[\varphi_0] \in \overline{K}$, $g[A] \cap K = \emptyset$ *and if Assumptions* H1, H2, H3* *and* H4* *hold then there exist a nonzero element* $q \in Z^*$, *an element* $\omega \in \Omega$ *and an integrable function* $E : [0, 1] \to B(X, Y)$ *with* $E[t] \in \Phi[t]$ *for all* $t \in [0, 1]$ *such that*

(1) $q(z - g[\varphi_0]) \geq 0$ *for all* $z \in K$,

(2) *the unique bounded integrable solution* $p : [0, 1] \to X^*$ *of the integral equation*

$$p[t] = \int_t^1 p[\tau] E[\tau] d\tau + q\omega([t, 1]) \quad \text{for every } t \in [0, 1]$$

will satisfy the condition

(3) $p[0] = 0$,

(4) $\int_0^1 p[t](f_0[\varphi_0[t], t] - f[\varphi_0[t], t]) dt \geq 0$ *for all* $f \in \Gamma$.

REMARK. If A is a convex subset of $B(X, Y)$ and B is a convex subset of $B(Y, Z)$ then the subset $BA \triangleq \{ba : b \in B, a \in A\}$ of $B(X, Z)$ is not necessarily convex. This implies in particular that the concept of shield does not lend itself to any form of chain rule. Since the optimal control problem considered in this paper contains compositions of functions there is a need for a second concept of set-valued derivative which would satisfy a chain rule and which would include the concept of shield as a particular case. I have shown in [17] that such a theory can be developed and I have given the name of *fan* to that second concept of set-valued derivative. The fan is a good working tool for proving theorems but the shield is a simpler way to state results. In [17] I show that Clarke's general derivative [3] is a shield and that the set of subgradients to a continuous convex function defined on a Banach space (Rockafellar [30]) is also a shield. The results of [17] can be considered as a calculus (including composition of functions) which allows us to combine usual derivative(s), Clarke's general derivative(s) and set(s) of subgradients of convex functions.

DEFINITION. If X and Y are Banach spaces, if $f : X \to Y$ then a subset A of $B(X, Y)$ is a *P-fan for* f *at* $x_0 \in X$ if A is strong operator compact and if for any compact set $\Omega \subset X$ and any $\varepsilon > 0$ there exists a $\delta > 0$ such that for any $\eta > 0$ there exists a strong operator continuous function $\alpha : \Omega \cap N(x_0, \delta) \to A$ such that $|f[x_0] + \alpha[x](x - x_0) - f[x]| \leq \eta + \varepsilon |x - x_0|$.

DEFINITION. If X and Y are Banach spaces, if $f : X \to Y$ then a subset A of $B(X, Y)$ is a *fan for* f *at* $x_0 \in X$ if $A = \bigcap_{i=1,2,\ldots} A_i$ where the sets

$A_1 \supset A_2 \supset \ldots$ are P-fans for f at x_0.

Lemmas 7.1, 7.2 and 7.3 which are stated below and proven in Halkin [17] are the key elements of the proof of Proposition 7.1.

LEMMA 7.1. *If X and Y are Banach spaces, $f : X \to Y$ is continuous in a neighborhood of x_0, $A \subset B(X, Y)$ is a shield for f at x_0, then A is a fan for f at x_0.*

LEMMA 7.2 (Chain rule for fans). *If X, Y and Z are Banach spaces, $f : X \to Y$ is continuous in a neighborhood of x_0, $g : Y \to Z$, $A \subset B(X, Y)$ is a fan for f at x_0, $B \subset B(Y, Z)$ is a fan for g at $f[x_0]$, then $BA \subset B(X, Z)$ is a fan for $g \circ f$ at x_0.*

LEMMA 7.3. *If X and Y are Banach spaces, $f : X \to Y$, $A \subset B(X, Y)$ is a fan for f at x_0, $\Omega \subset X$ is convex, $x_0 \in \overline{\Omega}$, $K \subset Y$ is convex and substantial, $f[x_0] \in \overline{K}$, for every $a \in A$ the sets $\{f[x_0] + a(x-x_0) : x \in \Omega\}$ and K are not separated, then there exists an $\eta > 0$ such that for any continuous function $g : \Omega \to Y$ with $|g[x] - f[x]| \leq \eta$ for all $x \in \Omega$ the sets $g[\Omega]$ and K are not disjoint.*

MULTIPLIER RULE FOR NONDIFFERENTIABLE MATHEMATICAL PROGRAMMING

PROBLEM. We are given Banach spaces X, Y_1 and Y_2, a finite-dimensional Euclidean space Y_3, a convex set $\Omega \subset X$, convex cones $K_1 \subset Y_1$ and $K_2 \subset Y_2$ with $0 \notin K_1$ and functions $f_i : X \to Y_i$ for $i = 1, 2$ and 3. Let $\tilde{A} = \{x : x \in \Omega, f_2[x] \in K_2, f_3[x] = 0\}$. An element $x_0 \in \tilde{A}$ is said to be optimal if for all $x \in \tilde{A}$ we have $f_1[x] - f_1[x_0] \notin K_1$. We assume that K_1 and K_2 have nonempty interiors, that f_i is continuous in a neighborhood of x_0 and admits $A_i \subset B(X, Y_i)$ as its fan at x_0 for $i = 1, 2$ and 3. Let $K_1^+ = \{f_1[x_0] + y_1 : y_1 \in K_1\}$ and let $K = K_1^+ \times K_2 \times K_3$. From Lemma 7.3 we obtain the following result.

PROPOSITION 7.2. *If x_0 is optimal then there exist a nonzero element $(\lambda_1, \lambda_2, \lambda_3) \in Y_1^* \times Y_2^* \times Y_3^*$ and an element $(a_1, a_2, a_3) \in A_1 \times A_2 \times A_3$ such that*

(1) $\sum_{i=1,2,3} \lambda_i a_i (x-x_0) \leq 0$ *for all* $x \in \Omega$,

(2) $\lambda_1 y_1 \geq 0$ *for all* $y_1 \in K_1$,

(3) $\lambda_2 y_2 \geq 0$ *for all* $y_2 \in K_2$,

(4) $\lambda_2 f_2[x_0] = 0$.

Appendix: Solutions of Integral Equations

In the present paper I need the following simple form of Gronwall's Inequality (see Hartman [18], for instance).

PROPOSITION A.1. *If* $\sigma : [a, b] \to [0, \infty)$ *is integrable,* $K \geq 0$, $r : [a, b] \to [0, \infty)$ *is continuous and*

$$r[t] \leq K + \int_a^t \sigma[\tau] r[\tau] d\tau \quad \text{for all} \quad t \in [a, b]$$

then

$$r[t] \leq K e^{\int_a^t \sigma[\tau] d\tau} \quad \text{for all} \quad t \in [a, b].$$

In Proposition A.2, I use Picard's Method and Gronwall's Inequality to prove the existence of a solution for the type of integral equations considered in the present paper. In Proposition A.3, I use again Gronwall's Inequality to show that the solution depends continuously on the data.

Let X be a finite dimensional Euclidean space and let $C(X)$ be the space of all continuous functions from $[0, 1]$ into X endowed with the sup norm. Let $\varphi_0 \in C[X]$, $\rho > 0$ and $\sigma : [0, 1] \to [0, \infty)$ be integrable. In Chapter 2, I have defined what is meant by a function $f : X \times [0, 1] \to X$ of class $\Xi(\rho, \sigma)$ around φ_0. Let us first remark that if $\varphi \in N(\varphi_0, \rho)$ and if $f : X \times [0, 1] \to X$ is of class $\Xi(\rho, \sigma)$ around φ_0, then the function $t \to f[\varphi[t], t] : [0, 1] \to X$ is integrable.

PROPOSITION A.2. *Let* $f : X \times [0, 1] \to X$ *be of class* $\Xi(\rho, \sigma)$ *around* φ_0, *let* $\psi \in C(X)$ *and let*

$$A = \sup_{t \in [0,1]} \left| \psi[t] + \int_0^t f[\varphi_0[\tau], \tau] d\tau - \varphi_0[t] \right|.$$

If $A < \rho e^{-\int_0^1 \sigma[\tau] d\tau}$ *then there exists a unique function* $\varphi \in C(X)$ *such that*

$$\varphi[t] = \psi[t] + \int_0^t f[\varphi[\tau], \tau] d\tau \quad \text{for all} \quad t \in [0, 1].$$

Moreover we have $\varphi \in N(\varphi_0, \rho)$.

Proof of Proposition A.2. Let $\eta > 0$ be such that

$$A = (\rho-\eta)e^{-\int_0^1 \sigma[\tau]d\tau}.$$

Let $k \in \{1, 2, \ldots\}$ such that

(1) $$\int_{i/k}^{(i+1)/k} \sigma[\tau]d\tau < \tfrac{1}{2}$$

and

(2) $$|\psi[t]-\psi[i/k]| + \int_{i/k}^{t} \sigma[\tau]d\tau + |\varphi_0[t]-\varphi_0[i/k]| < \eta$$

for all $i \in \{0, 1, \ldots, k-1\}$ and all $t \in [i/k, (i+1)/k]$.

For every $i \in \{0, 1, \ldots, k\}$ I shall denote by P_i the following statement: there exists a unique continuous function φ from $[0, i/k]$ into X such that

(3) $$\varphi[t] = \psi[t] + \int_0^t f[\varphi[\tau], \tau]d\tau$$

and

(4) $$|\varphi[t]-\varphi_0[t]| \leq \rho - \eta$$

for all $t \in [0, i/k]$.

We see that P_0 is trivially true and that P_k is the desired result. I shall prove Proposition A.2 by showing that P_i implies P_{i+1} for all $i \in \{0, \ldots, k-1\}$.

Let $i \in \{0, \ldots, k-1\}$ and let us assume that P_i is true. Let K be the set of all continuous functions $\emptyset$ from $[i/k, (i+1)/k]$ into X such that $|\emptyset[t]-\varphi_0[t]| \leq \rho$ for all $t \in [i/k, (i+1)/k]$. We shall use the sup norm $|\cdot|$ on K. Let $\emptyset \to T_\emptyset$ be defined by

$$T_\emptyset[t] = \varphi[i/k] + \psi[t] - \psi[i/k] + \int_{i/k}^t f[\emptyset[\tau], \tau]d\tau$$

for all $\emptyset \in K$ and all $t \in [i/k, (i+1)/k]$. From our assumptions it follows that $T_\emptyset \in K$ for all $\emptyset \in K$ and that

$$|T_{\emptyset_2}-T_{\emptyset_1}| \leq \tfrac{1}{2}|\emptyset_2-\emptyset_1|$$

whenever $\emptyset_1, \emptyset_2 \in K$. Let $\hat{\emptyset}$ be the unique fixed point of T. We shall extend the definition of φ by the relation

$$\varphi[t] = \hat{\emptyset}[t] \text{ for all } t \in [i/k, (i+1)/k].$$

The function φ is now defined on the set $[0, (i+1)/k]$ where it satisfies condition (3), since $\hat{\varphi}$ is a fixed point of T. For all $t \in [0, (i+1)/k]$ we have

$$|\varphi[t]-\varphi_0[t]| \leq A + \int_0^t |f[\varphi[\tau], \tau]-f[\varphi_0[\tau], \tau]|d\tau$$

$$\leq A + \int_0^t \sigma[\tau]|\varphi[\tau]-\varphi_0[\tau]|d\tau ,$$

and hence, according to Proposition A.1,

$$|\varphi[t]-\varphi_0[t]| \leq Ae^{\int_0^1 \sigma[\tau]d\tau} = \rho - \eta .$$

The function φ satisfies condition (4) on the set $[0, (i+1)/k]$. This concludes the proof of Proposition A.2.

PROPOSITION A.3. *Let* $f_1, f_2 : X \times [0, 1] \to X$ *be of class* $\Xi(\rho, \sigma)$ *around* φ_0, *let* $\psi_1, \psi_2 \in C(X)$ *and let* $\varphi_1, \varphi_2 \in N(\varphi_0, \rho)$ *such that for all* $i \in \{1, 2\}$ *and* $t \in [0, 1]$ *we have*

$$\varphi_i[t] = \psi_i[t] + \int_0^t f_i[\varphi_i[\tau], \tau]d\tau .$$

If

$$B = \sup_{t \in [0,1]} \left|\psi_2[t]-\psi_1[t] + \int_0^t (f_2[\varphi_1[\tau], \tau]-f_1[\varphi_1[\tau], \tau])d\tau\right|$$

then $|\varphi_2-\varphi_1| \leq Be^{\int_0^1 \sigma[\tau]d\tau}$.

Proof of Proposition A.3. For all $t \in [0, 1]$ we have

$$|\varphi_2[t]-\varphi_1[t]| \leq B + \int_0^t |f_2[\varphi_2[\tau], \tau]-f_2[\varphi_1[\tau], \tau]|d\tau$$

$$\leq B + \int_0^t \sigma[\tau]|\varphi_2[\tau]-\varphi_1[\tau]|d\tau$$

and hence, according to Proposition A.1, we have

$$|\varphi_2-\varphi_1| \leq Be^{\int_0^1 \sigma[\tau]d\tau}$$

This concludes the proof of Proposition A.3.

References

[1] Michael D. Canon, Clifton D. Cullum, Jr., Elijah Polak, *Theory of Optimal Control and Mathematical Programming* (McGraw-Hill, New York, 1970).

[2] C. Carathéodory, *Calculus of Variations and Partial Differential Equations of the First Order*, Part II: *Calculus of Variations* (Holden-Day Series in Mathematical Physics. Holden-Day, San Francisco, Cambridge, London, Amsterdam, 1967).

[3] Frank H. Clarke, "Generalized gradients and applications", *Trans. Amer. Math. Soc.* **205** (1975), 247-262.

[4] Nelson Dunford and Jacob T. Schwartz, *Linear Operators*. Part I: *General Theory* (Pure and Applied Mathematics, **7**. Interscience, New York, London, 1958).

[5] I.V. Girsanov, *Lectures on Mathematical Theory of Extremum Problems* (Lecture Notes in Economics and Mathematical Systems, **67**. Springer-Verlag, Berlin, Heidelberg, New York, 1972).

[6] Hubert Halkin, "On the necessary condition for optimal control of nonlinear systems", *J. Analyse Math.* **12** (1964), 1-82.

[7] Hubert Halkin, "An abstract framework for the theory of process optimization", *Bull. Amer. Math. Soc.* **72** (1966), 677-678.

[8] Hubert Halkin, "Convexity and control theory", *Functional Analysis and Optimization*, 85-97 (Academic Press, New York, London, 1966).

[9] Hubert Halkin, "A property of nonseparated convex sets", *Proc. Amer. Math. Soc.* **17** (1966), 1389-1395.

[10] Hubert Halkin, "Nonlinear nonconvex programming in an infinite dimensional space", *Mathematical Theory of Control*, 10-25 (Academic Press, New York, London, 1967).

[11] H. Halkin, "A satisfactory treatment of equality and operator constraints in the Dubovitskii-Milyutin optimization formalism", *J. Optimization Theory Appl.* **6** (1970), 138-149.

[12] Hubert Halkin, "Implicit functions and optimization problems without continuous differentiability of the data", *SIAM J. Control* **12** (1974), 229-236.

[13] Hubert Halkin, "Necessary conditions in mathematical programming and optimal control theory", *Optimal Control Theory and its Applications* (Lecture Notes in Economics and Mathematical Systems, **105**, 113-165. Springer-Verlag, Berlin, Heidelberg, New York, 1974).

[14] Hubert Halkin, "Brouwer fixed point theorem versus contraction mapping theorem in optimal control theory", *International Conference on Differential Equations*, 337-344 (Academic Press Rapid Manuscript Reproduction. Academic Press [Harcourt Brace Jovanovich], New York, San Francisco, London, 1975).

[15] Hubert Halkin, "Interior mapping theorem with set-valued derivatives", *J. Analyse Math.* **30** (1976), 200-207.

[16] Hubert Halkin, "Mathematical programming without differentiability", *Calculus of Variations and Control Theory*, 279-287 (Academic Press Rapid Manuscript Reproduction. Academic Press, New York, San Francisco, London, 1976).

[17] Hubert Halkin, "Optimization without differentiability", submitted.

[18] Philip Hartman, *Ordinary Differential Equations* (John Wiley & Sons, New York, London, Sydney, 1964. Reprinted: Baltimore, 1973).

[19] Magnus R. Hestenes, *Calculus of Variations and Optimal Control Theory* (John Wiley & Sons, New York, London, Sydney, 1966).

[20] Fritz John, "Extremum problems with inequalities as subsidiary conditions", *Studies and Essays* (Courant Anniversary Volume, 187-204. Interscience, New York, 1948).

[21] E.B. Lee and L. Markus, *Foundations of Optimal Control Theory* (The SIAM Series in Applied Mathematics. John Wiley & Sons, New York, London, Sydney, 1967).

[22] Olvi L. Mangasarian, *Nonlinear Programming* (McGraw-Hill, New York, London, Sydney, 1969).

[23] E.J. McShane, "On multipliers of Lagrange problems", *Amer. J. Math.* **61** (1939), 809-819.

[24] Lucien W. Neustadt, *Optimization: A Theory of Necessary Conditions* (Princeton University Press, Princeton, New Jersey, 1976).

[25] L.S. Pontryagin, V.G. Boltyanskii, R.V. Gamkrelidze, and E.F. Mishchenko, *The Mathematical Theory of Optimal Processes* (Interscience [John Wiley & Sons], New York, London, 1962).

[26] Bruce Hunter Pourciau, "A generalized derivative and its applications to optimization theory, classical analysis, and global univalence" (PhD Dissertation, University of California at San Diego, San Diego, 1976).

[27] B.H. Pourciau, "Analysis and optimization of Lipschitz continuous maps", *J. Optimization Theory Appl.* **22** (1977), 311-352.

[28] B.H. Pourciau, "Global univalence of Lipschitz continuous mappings", submitted.

[29] B.N. Pshenichnyi, *Necessary Conditions for an Extremum* (Pure and Applied Mathematics, **4**. Marcel Dekker, New York, 1971).

[30] R.T. Rockafellar, "Characterization of the subdifferentials of convex functions", *Pacific J. Math.* **17** (1966), 497-510.

[31] T.H. Sweetser, III, "A minimal set-valued strong derivative for vector valued Lipschitz functions", *J. Optimization Theory Appl.* (to appear).

[32] J. Warga, "Necessary conditions for minimum in relaxed variational problems", *J. Math. Anal. Appl.* **4** (1962), 129-145.

[33] J. Warga, "Necessary conditions without differentiability in optimal control", *J. Differential Equations* (to appear).

[34] Sadayuki Yamamuro, *Differential Calculus in Topological Linear Spaces* (Lecture Notes in Mathematics, **374**. Springer-Verlag, Berlin, Heidelberg, New York, 1974).

GENERAL CONTROL SYSTEMS

P.E. Kloeden

1. Abstract dynamical systems

Attainability sets $F(x_0, t_0, t)$ play a fundamental role in the theory of dynamical systems without uniqueness, notably control systems. They provide a starting point for the development of an abstract, axiomatic approach to such systems, which enables concepts and different modes of behaviour to be investigated in some generality without their inherent features being obscured by circumstantial details pertaining to a particular function or analytical representation.

The axiomatic scheme for dynamical systems with uniqueness is well known [6, 17], but is worth briefly reviewing here as it gives some insight into what is required in its generalisation to dynamical systems without uniqueness. It draws its initial motivation from autonomous ordinary differential equations

$$\dot{x} = f(x) \qquad (1.1)$$

on a phase or state space $X = \mathbb{R}^n$. If uniqueness, global existence and global extension are assumed for the initial value problem for this equation, its solutions can be represented by a single-valued mapping

$$\pi : X \times \mathbb{R} \to X$$

which satisfies the following four properties (for example, see [41]):

I. π *is defined for all* $(x, t) \in X \times \mathbb{R}$ *(global definition and global extension;*

II. $\pi(x, 0) = x$ *for all* $x \in \mathbb{R}$ *(initial condition);*

III. $\pi\bigl(\pi(x, s), t\bigr) = \pi(x, t+s)$ *for all* $x \in X$ *and* $s, t \in \mathbb{R}$ *(group evolution);*

IV. π *is jointly continuous in* (x, t) *for all* $(x, t) \in X \times \mathbb{R}$ *(continuity).*

The group evolution property III here is quite significant and is a consequence of the evolution of the system depending only on the present state of the system, not on how the present state was attained. It can be thought of as a degenerate Markovian property. Other properties, such as differentiability, are also satisfied by the solutions of (1.1) but are not considered as fundamental as properties I-IV.

Abstract dynamical systems with uniqueness are then defined axiomatically in terms of such mappings π which satisfy properties I-IV. Usually also X is taken to be a general topological space and sometimes the time set $\mathbb{R}$ is replaced by an additive topological group T. The resultant theory of such systems has become known as *topological dynamics* (for example, [17, 59]), or in the special case where X is metric and $T = \mathbb{R}$ as *dynamical systems theory* (for example, [6]), and includes, but is not just restricted to, differential systems such as (1.1).

The scope of such a theory is however restricted by the autonomy, global definition in x, global extension in t and uniqueness or single-valuedness of the mapping π [40]. The first three of these can be readily relaxed without major conceptual change: autonomy by simple adding another variable; global definition and global extension by restricting the domain of π to a subset of $X \times \mathbb{R}$. The latter gives rise to *local-semi-dynamical systems* (for example, [5]), or more specifically when the domain of π is restricted to $X \times \mathbb{R}^+$ to *semi-dynamical systems*, and broadens the theory to include Markov processes on spaces of probability measures and the solutions of functional differential equations. Relaxation of uniqueness can be done in one of two ways, by using either sets of trajectories or attainability sets. Both of these require major formal changes to the mapping π.

2. Dynamical systems without uniqueness

Two typical and important examples of dynamical systems without uniqueness are *ordinary differential equations without uniqueness* ([32, 58] for example $\dot{x} = x^{2/3}$, with its nonlipschitzian right-hand side) and *ordinary differential control systems* ([36], that is, $\dot{x} = f(x, t, u)$ where $u \in U \subset \mathbb{R}^m$). Both are also special cases of *contingent equations* ([8, 9, 48], that is, $\dot{x} \in F(x, t) \subset \mathbb{R}^n$). The control systems have more significant applications and thus provide greater motivation for studying dynamical systems without uniqueness, though historically the initial motivation came from ordinary differential equations without uniqueness.

Given such a differential system and an initial condition $x(t_0) = x_0$, the usual procedure is to find all of its trajectories satisfying this initial condition, the set of which will be denoted by $\Phi(x_0, t_0)$, and then to define the attainability sets $F(x_0, t_0, t)$ in terms of these trajectories, namely

$$F(x_0, t_0, t) = \cup\{x \in X; \; x = \phi(t) \text{ for some } \phi \in \Phi(x_0, t_0)\}.$$

This procedure gives the impression that trajectories are the fundamental entities in a dynamical system without uniqueness and that an abstract, axiomatic theory of such systems should be based on them. This indeed was the viewpoint first adopted in 1932 by Fukuhara [16], in 1937 by Zaremba [70] and, more recently, in 1963 by Bushaw [14]. Bushaw introduced the apt name *dynamical polysystems* and, unlike his predecessors, also considered control theoretic questions. Far more popular, however, has been the use of attainability sets as the starting point in an axiomatic treatment of dynamical polysystems, inspite of their being defined on the differential level in terms of the more basic trajectories. The reason for this is that the dynamical and geometric character of such systems is more transparent with attainability sets than with sets of trajectories. Also attainability sets provide a natural generalisation of dynamical systems with uniqueness, from single-valued mappings π to set-valued attainability mappings F.

The basic properties of attainability sets of ordinary differential equations without uniqueness have been known for a long time, the most famous, their compactness and connectedness being established by Kneser in 1923 [32, 58]. Abstract dynamical polysystems were first defined axiomatically in terms of attainability sets with such properties by Barbašin [2, 3] and Minkevič [37-39] in the 1940's and extended to include control systems and contingent equations by Roxin [45-52] in the 1960's. These systems were defined on metric state spaces and called *general dynamical systems* or *general control systems*. Further generalisations have since been made to more general topological state spaces and to attainability relations without mention of topologies [10, 11, 15, 18, 40], but the most far reaching results obtained have been for general dynamical and general control systems on metric state spaces, so attention will be restricted to them in the sequel.

3. General dynamical systems

Barbašin [2] initiated the study of general dynamical systems in 1948 using axioms based on the properties of the attainability sets of autonomous ordinary differential equations without uniqueness. He defined a *general dynamical system* on a metric state space X by means of a set-valued attainability mapping

$$F : X \times \mathbb{R} \to X$$

satisfying the following axioms:

I. $F(x, t)$ *is a nonempty, compact connected subset of* X *for all* $(x, t) \in X \times \mathbb{R}^+$ *(global definition, infinite forwards extendability, Kneser's theorem);*

II. $F(x, 0) = \{x\}$ *for all* $x \in X$ *(initial condition);*

III. $F(x, s+t) = F(F(x, s), t) = \cup\{F(y, t); y \in F(x, s)\}$ *for all* $x \in X$ *and* $s, t \geq 0$ *(semigroup evolution);*

IV. $F(x, t)$ *is continuous in* t *with respect to the Hausdorff metric for each fixed* $x \in X$ *(continuity in time), that is,*

$$\lim_{s \to t} \rho\bigl(F(x, s), F(x, t)\bigr) = 0 ;$$

V. $F(x, t)$ *is upper semicontinuous in* (x, t) *with respect to the Hausdorff metric (in effect, upper semicontinuity in initial conditions), that is,*

$$\lim_{\substack{y \to x \\ s \to t}} \rho^*\bigl(F(y, s), F(x, t)\bigr) = 0 ;$$

VI. $F(x, t)$ *satisfies the same properties for* $t \leq 0$ *as for* $t \geq 0$ *(backwards extendability).*

The *Hausdorff metric* ρ on compact subsets of X (with metric d) used above is defined as

$$\rho(A, B) = \max\{\rho^*(A, B), \rho^*(B, A)\}$$

where

$$\rho^*(A, B) = \max\{\rho(a, B); a \in A\}$$

and

$$\rho(a, B) = \min\{d(a, b); b \in B\} .$$

Axioms I-VI above clearly include the axioms of a dynamical system with uniqueness π and in fact reduce to them when the attainability sets of F are single-valued everywhere. The only one of them which is, possibly, of any surprise is axiom V on upper semicontinuity in initial conditions. The following example of an ordinary differential equation without uniqueness shows that it is by no means a pathological phenomenon and that it cannot be strengthened and amalgamated with axiom IV to continuity in (x, t) without excluding a large class of simple and common systems.

EXAMPLE 1. For $X = \mathbb{R}$ the ordinary differential equation

$$\dot{x} = \frac{1}{3} x^{2/3}$$

has nonunique solutions for $x = 0$ only. By elementary integration its attainability sets are found to be

$$F(x, t) = \begin{cases} \{(x^{1/3}+t)^3\} & \text{for } x \neq 0, \\ [0, t^3] & \text{for } x = 0. \end{cases}$$

Hence, for $y \to 0+$ and $s \to t > 0$,

$$F(y, s) = \{(y^{1/3}+s)^3\} \to \{t^3\} \subsetneq [0, t^3] = F(0, t),$$

so F is upper semicontinuous, but not continuous, in (x, t) at $(0, t)$.

General dynamical systems, or *dispersive dynamical systems* as they have sometimes been called, received some attention in the Soviet literature [3, 12, 13, 37, 38, 39, 71] for a few years following the publication of Barbašin's paper [2] in 1948, but interest then waned until the late 1960's when it was revived by Izman [19-22] and Šibirskii, Čeban and Čirkova [60-62] from Kishinev in the USSR and by Szegö and Treccani [65-67] from Milan in Italy. These later workers attempted to generalise the behavioural classification of dynamical systems with uniqueness [6, 17, 59] to general dynamical systems. More will be said on this in Section 6.

It should be noted that although not demanded in the definition of a general dynamical system most of the significant results in the above papers required the state space X to be *complete* and *locally compact*, the latter of which being the most severe restriction in the theory of general dynamical systems, and also general control systems, as developed to date.

4. General control systems

From the viewpoint of applications of the underlying systems, the most significant development in the theory of dynamical polysystems occurred in the 1960's when Roxin extended Barbašin's work on general dynamical systems to include control systems governed by ordinary differential equations and, more generally, contingent equations [45-48]. Roxin called these systems *general control systems*. They are defined on a complete locally compact metric state space X and satisfy slightly more general axioms than axioms I-VI above used by Barbašin. Roxin took general control systems to be nonautonomous and relaxed axiom I to:

$F(x_0, t_0, t)$ *is a nonempty, closed subset of* X *for each* $x_0 \in X$ *and all* $t_0 \leq t$.

Both compactness and connectedness have been omitted here, but Roxin showed [47, Theorem 4.2] that compactness follows from axioms II and V and from the local compactness of X. He also showed [47, Example 4.1] that connectedness is independent of the other axioms.

The remaining change by Roxin to Barbašin's axioms was to relax the backwards extendability axiom VI to:

for all $x_1 \in X$ and $t_0 < t_1$ there exists an $x_0 \in X$ such that $x_1 \in F(x_0, t_0, t_1)$;

that is, any given state can be attained from some other state over any given time interval. From this the backwards extension of $F(x_0, t_1, t)$ can be defined for $t \leq t_0$ as the set of all states from which x_0 can be attained at time t_0 starting from time t. Further it satisfies all of the axioms as for $t \geq t_0$, except possibly continuity in t, and hence compactness [47, Section 5]. Consequently the above statement is slightly weaker than Barbašin's axiom VI. A justification for it is that in a control context the future evolution of the system rather than its past is of predominant interest.

Actually, as noted by Roxin in [47], the attainability sets of most nice and frequently used control systems satisfy stronger properties than those used as axioms of general control system. These axioms however allow "not so nice" control systems, as well as ordinary differential equations without uniqueness and contingent equations to also be included in the theory.

The axioms of general control systems have been further relaxed in three directions: Roxin took the restriction of a general control system to a closed subset Y of the state space X to obtain a *local general control system* [50]; de Blasi and Schinas [7] and Szegö and Treccani [67] considered *discrete-time general dynamical systems*; and Kloeden omitted the backwards extendability axiom altogether to obtain *general semidynamical systems* [24-26, 31]. This omission of the backwards extendability axiom requires only minor, but inconvenient, modifications to most of Roxin's proofs and considerably extends the scope of the theory, notably to include stochastic control systems on compact spaces of probability measures arising from stochastic differential equations and Markov transition matrices [26, 56]. The requirement of local compactness of the state space however excludes many important systems on function state spaces arising from functional and partial differential equations [33, 42].

A significant and convenient feature of general control systems is that there is no explicit mention of control functions, which results in greater generality and considerable technical convenience. The role of such control functions is taken over by the trajectories of a general control system, which are defined in terms of its attainability sets.

5. Trajectories of general control systems

The attainability sets of a differential system are usually defined in terms of its trajectories, but in the definition of general control systems they are specified without mention of trajectories. For this latter approach to be intuitively consistent trajectories should be reclaimable from the attainability sets and each point in an attainability set should actually be attainable by such a trajectory.

Following Barbašin [2], Roxin [47] defined a *trajectory* of a general control system F as a single-valued mapping $\phi : [t_0, t_1] \to X$ such that $\phi(t) \in F(\phi(s), s, t)$ for all $t_0 \leq s \leq t \leq t_1$. See also [66]. Continuity of ϕ is not assumed in advance here, but follows from axioms II and V of a general control system [47, Lemma 6.1]. The question of existence of trajectories and the attainability by such trajectories of points in the attainability sets is closely related to that of existence of continuous selectors [43] and is answered affirmatively in the following theorem of Barbašin [2] (see [47, Theorem 6.1] for general control systems and [25, Theorem 4.2] for general semidynamical systems).

THEOREM. *If, for a certain general control system* F, $x_1 \in F(x_0, t_0, t_1)$, *then there exists a trajectory* $\phi : [t_0, t_1] \to X$ *of* F *such that* $\phi(t_0) = x_0$ *and* $\phi(t_1) = x_1$.

This theorem is proved by actually constructing a trajectory ϕ satisfying the given boundary conditions. This can be done without loss of generality for the unit time interval. The first step is to set $\phi(0) = x_0$ and $\phi(1) = x_1$, and then to define ϕ successively for the dyadic fractions $1/2, 1/4, 3/4, \ldots, p/2^q$ as follows: choose $\phi(1/2)$ to be any point in the nonempty intersection

$$F(\phi(0), 0, 1/2) \cap F(\phi(1), 1, 1/2).$$

This procedure is then repeated with the unit interval $[0, 1]$ replaced by the half intervals $[0, 1/2]$ and $[1/2, 1]$ to choose $\phi(1/4)$ and $\phi(3/4)$, respectively, and so on for the remaining values of $\phi(p/2^q)$. The second step is to define $\phi(t)$ at non-dyadic instants t by the limiting process

$$\phi(t) \in \cap \{F(\phi(t_i), t_i, t) \cap F(\phi(t_j), t_j, t); t_i < t < t_j, t_i \text{ and } t_j \text{ dyadic}\}.$$

When a differential system underlies a general control system the above trajectories correspond to generalised solutions in the case of ordinary differential equations without uniqueness [66, Theorem 5.1] and to the solutions of relaxed controls in the case of ordinary differential control systems.

Another important theorem of Barbašin (see [47, Theorem 6.2]) says that the set $\Phi(x_0, t_0; F)$ of all trajectories of a general control system F with $\phi(t_0) = x_0$ is sequentially compact with respect to uniform convergence on compact time sets. Generalisations of this theorem are given in [25, 28].

In general the above sets $\Phi(x_0, t_0; F)$ of trajectories contain more than one trajectory. This lead Roxin [45-47, 49, 51, 52] to distinguish two versions of any given behavioural property, depending on whether every trajectory (that is, the entire attainability funnel) or just some trajectories emanating from each initial state in question satisfied the desired property, for example, invariance, asymptotic invariance, attraction, Lyapunov stability. He called such versions of a given property *strong* and *weak*, respectively. The strong versions are of main interest for systems arising from ordinary differential equations without uniqueness and have been extensively investigated by the Soviet and Italian workers in the subject. The weak versions are of greater significance in a control system in which not every trajectory need satisfy a given property and for which there exists a mechanism, namely the control functions, for selecting particular trajectories. They have been investigated mainly by Roxin [45-47, 49, 51, 52] and Kloeden [27-29].

6. Lyapunov stability theory for general control systems

There have been many variations to the basic concept of Lyapunov stability of an equilibrium point or, more generally, of an invariant set with respect to a dynamical system [69]. All, however, share the common feature that small deviations from the invariant set result in only small changes in the subsequent behaviour of the system and that stability is usually tested for indirectly with the aid of an energy-like Lyapunov function. On the differential level this use of Lyapunov functions is a great advantage as it does not require explicit knowledge of the trajectories of the system, but is counterbalanced by the difficulty in finding a suitable Lyapunov function.

A lot of work has been done in deriving necessary and sufficient conditions involving Lyapunov functions for these stabilities. These derivations are usually easier and the conditions more widely applicable when the dynamics are described by general control systems [19-22, 27-29, 47, 49, 51, 52, 57, 65-67, 71]. In addition geometric character of general control systems often make it easier to distinguish and compare different types of stabilities [49].

The following examples of strong and weak stability of a subset A of the state space X relative to a general control system are representative of Lyapunov stabilities defined in terms of general control systems. The strong version is due to Zubov [71] and the weak version to Roxin [47].

DEFINITION. *Relative to a general control system F a subset A of X is*
(i) *strongly stable if for every* $\varepsilon > 0$ *and* $t_0 \in R$ *there exists a*
$\delta = \delta(\varepsilon, t_0) > 0$ *such that for all* $t \geq t_0$,

$$F(S_\delta(A), t_0, t) \subset S_\varepsilon(A) ;$$

(ii) weakly stable *if for every* $\varepsilon > 0$ *and* $t_0 \in \mathbb{R}$ *there exists a*
$\delta = \delta(\varepsilon, t_0) > 0$ *such that*

$$\phi(t) \in S_\varepsilon(A)$$

for all $t \geq t_0$ *and some* $\phi \in \Phi(x_0, t_0; F)$ *for each* $x_0 \in S_\delta(A)$.

In this definition $S_\eta(A)$ denotes the η-neighbourhood of the set A ; that is

$$S_\eta(A) = \cup\{x \in X; \rho(x, A) < \eta\} \; .$$

Clearly, a strongly stable set is also weakly stable. The converse is in general not true. Also, a strongly stable set is *strongly invariant*, whereas a weakly stable set need only be *weakly invariant* [47, Section 7].

The following theorem is due to Zubov [71] for strong stability and Roxin [47] for weak stability. It is often useful to imagine the Lyapunov function $V(x, t)$ in this theorem as a potential energy function, in which case the conditions of the theorem can be interpreted as there being a potential energy trough about the set $A \times \mathbb{R}$ such that the potential energy decreases along trajectories (all in strong stability) of the general control system.

THEOREM. *A closed set* A *is strongly stable relative to a general control system* F *if and only if there exists a real valued function* $V(x, t)$ *such that:*

(1) $V(x, t)$ *is defined for all* $(x, t) \in S_\eta(A) \times \mathbb{R}$ *for some* $\eta > 0$;

(2) $V(x, t) = 0$ *for all* $(x, t) \in A \times \mathbb{R}$;

(3) $a(\rho(x, A)) \leq V(x, t) \leq b(\rho(x, A), t)$ *for all* $(x, t) \in S_\eta(A) \times \mathbb{R}$,
where $a(r)$ *is a real valued, continuous, strictly increasing function of* r *with* $a(0) = 0$, *and similarly for* $b(r, t)$ *for each fixed* t ;

(4) $D^*V(x, t) \leq 0$ *for all* $(x, t) \in S_\eta(A) \times \mathbb{R}$.

For weak stability the conditions differ only in that $V(x, t)$ must be lower semicontinuous and the derivative $D^*V(x, t)$ in condition (4) is replaced by $D_+V(x, t)$.

In this theorem $D^*V(x, t) = \max\{D^+V(x, t), D^-V(x, t)\}$, where $D^\pm V(x, t)$ and $D_+V(x, t)$ are *generalised Dini derivatives* of V at (x, t) relative to the general control system F , defined as [47, Section 8]:

$$D^\pm V(x, t) = \varlimsup_{s \to t\pm} \left\{ \frac{V(y, x) - V(x, t)}{s - t}; y \in F(x, t, s) \right\}$$

and

$$D_+V(x, t) = \lim_{s \to t+} \inf\left\{\frac{V(y,s)-V(x,t)}{s-t}; y \in F(x, t, s)\right\}.$$

Often condition (4) of the theorem involving these generalised derivatives can be replaced by equivalent non-differential conditions [40, 47].

The definitions and theorems above typify Lyapunov stability theory for general control systems, in particular in regard to the difference in the necessary and sufficient conditions for strong and weak stabilities. Many other types of stability have been considered, including uniform, asymptotic, finite-time, eventual and relative stabilities. See [19-22, 57, 65-66, 71] for strong versions, [27-29, 47, 49, 51, 52] for both strong and weak versions, and [7, 67] for discrete-time versions of such stabilities. Also Lagrange and Poisson stabilities have been considered in [11, 60] and boundedness of general control systems in [64]. In most of these papers the local compactness of the state space is crucial in the derivation of necessary and sufficient conditions.

7. Topologies for general control systems

Attention has so far been concentrated on a single general control system at a time, but will now be turned to the entire class F of all general control systems on a state space X, in particular to the various ways in which this class can be topologised. If a general control system is regarded as a single-valued mapping from $Y (= X \times \mathbb{R} \times \mathbb{R})$ into $\mathbb{K}$ (equals the space of all nonempty, compact subsets of X) this reduces to an exercise in function space topology [1, 23], but is by no means straight-forward on account of the mixed continuity axioms IV and V of general control systems. A detailed account of what follows can be found in [31]. Again the local compactness of X (and hence of Y) plays a crucial role here too.

In view of the mixed continuity of general control systems, two topologies will be needed for $\mathbb{K}$, the *Hausdorff metric topology* H and *Ponomarev topology* P [44]. Since X is a locally compact metric space, the P-topology can be generated by a neighbourhood system of the form

$$\{A \in \mathbb{K}; \rho^*(A, A_0) < \varepsilon\}$$

for all $A_0 \in \mathbb{K}$ and all $\varepsilon > 0$. Consequently the P-topology here is coarser than the H-topology. Indeed upper-semicontinuity with respect to the Hausdorff metric (that is, axiom V of a general control system) of a mapping $F : Y \to \mathbb{K}$ is equivalent to continuity with respect to the P-topology on $\mathbb{K}$ [24, Theorem 5.3.1].

The simplest useful topology for F is the topology T_1 of *pointwise convergence* in $(\mathbb{K}, H)$. It is generated by subbasic sets of the form

$$\{F \in F; F(y) \in U\}$$

for all $y \in Y$ and all H-open subsets $\mathbb{U}$ of $\mathbb{K}$. Convergence of a net of general control systems $F_\nu \to F$ in $(\mathbb{F}, T_1)$ is, as the name suggests, equivalent to pointwise convergence in $(\mathbb{K}, H)$; that is,

$$(\lambda_1) \qquad \rho(F_\nu(y), F(y)) \to 0 \quad \text{for all } y \in Y.$$

Since $(\mathbb{K}, H)$ is a metric space, T_1 is a Hausdorff (that is T_1-) topology. In fact it is the topology of a uniformity, but for many purposes is much too coarse a topology. What usually is required is a topology corresponding to continuous convergence, but in view of the mixed continuity properties of general control systems, it is not immediately apparent what continuous convergence is for them.

For general control systems which are continuous in *all* variables with respect to the Hausdorff metric the continuous convergence of a net $F_\nu \to F$ is simply continuous convergence of H-continuous mappings from Y into $\mathbb{K}$; that is,

$$(\lambda_2) \qquad \rho(F_\nu(y_\nu), F(y)) \to 0 \quad \text{for all nets } y_\nu \to y \text{ in } Y.$$

Since Y is locally compact, by a theorem of Arens [1, Theorem 4] the corresponding topology T_2 is the *compact-open* topology for H-continuous mappings from Y into $\mathbb{K}$ and is generated by subbasic sets of the form

$$\{F; F \text{ is } H\text{-cts and } F(y) \in \mathbb{U} \text{ for all } y \in A\}$$

for all H-open subsets $\mathbb{U}$ of $\mathbb{K}$ and all compact subsets A of Y.

The fact that λ_2-convergence of a constant net $F_\nu \equiv F$ reduces to a statement of the H-continuity of F suggests that the analogous definition of continuous convergence of general control systems in $\mathbb{F}$ cannot be expressed as a single convergence, but rather must be expressed as the conjunction of two convergences, one for each of the continuity axioms IV and V, namely,

$$(\lambda_3) \qquad \rho(F_\nu(x, t, s_\nu), F(x, t, s)) \to 0 \quad \text{for all nets}$$
$$(x, t, s_\nu) \to (x, t, s) \in Y \text{ which are constant in their first two}$$
co-ordinates, and

$$(\lambda_4) \qquad \rho^*(F_\nu(y_\nu), F(y)) \to 0 \quad \text{for all nets } y_\nu \to y \text{ in } Y.$$

The first of these, λ_3, corresponds to convergence in the topology T_3 which is a modification of the compact-open topology T_2 above and is generated by subbasic sets of the form

$$\{F \in \mathbb{F}; F(y) \in \mathbb{U} \text{ for all } y \in A\}$$

for all H-open subsets $\mathbb{U}$ of $\mathbb{K}$ and all compact subsets A of Y of the form

$\{x\} \times \{t\} \times B$, where B is a compact subset of $\mathbb{R}$. The second convergence, λ_4, is just continuous convergence of P-continuous mappings from Y into $\mathbb{K}$ and thus corresponds to the compact-open topology T_4 generated by subbasic sets

$$\{F \in F; F(y) \in \mathbb{U} \text{ for all } y \in A\}$$

for all P-open subsets $\mathbb{U}$ of $\mathbb{K}$ and all compact subsets A of Y.

Hence continuous convergence $\lambda_3 \cap \lambda_4$ of general control systems corresponds to convergence in the union topology $T_3 \cup T_4$ on F. This is a Hausdorff topology and is usually sufficient for most purposes.

An interesting situation arises when F is considered as a subset of a larger class G of mappings from Y into $\mathbb{K}$ and the topology, now defined on G, is required to be such that F is a *closed* subset of G, that is, such that the axioms of a general control system are preserved in the limit of a net of general control systems convergent under this topology. It is well known from real analysis that to ensure the continuity of a limit of a convergent net of continuous mappings, the convergence used must be the conjunction of continuous convergence and *uniform convergence on compacta*. This also holds for general control systems, but only for those general control systems which are H-continuous in all variables is the limit a general control system (also H-continuous in all variables). Example 1 of [31] shows that otherwise the limit need not satisfy the semigroup property, axiom III, of a general control system. To ensure that this property is also satisfied, and hence that the limit is a general control system, the convergence must further be strengthened with the inclusion of a *semigroup preserving convergence* (which is really just a ($\mathbb{K}$, H)-pointwise convergence for another class of mappings which is the homeomorphic image of G). Details of this can be found in [31].

8. Game dynamics and general control systems

The fundamental objective in mathematical game theory is to investigate and develop ways of resolving conflict situations. With differential games this is often made quite difficult by heavy analytical demands of the differential formulation of game dynamics, which, while mathematically necessary and interesting, are really peripheral to the task at hand. An abstract, axiomatic formulation of game dynamics offers a way of overcoming, or at least circumventing, such difficulties and also of obtaining deeper insight into the conflict situation.

The earliest use of axiomatically defined game dynamics appears to have been by Ryll-Nardzewski [55] for pursuit-evasion games, with each players dynamics being specified by sets of trajectories. Quite similar was the paper [68] by Varaiya, which was also for pursuit-evasion games, but with each player's dynamics described by a general control system. A characteristic feature of pursuit-evasion games is that each player controls a distinct dynamical system uncoupled from the one controlled by the other player. A more interesting situation occurs when both players have some, but not complete control over the same dynamical system. General control systems have been used by Roxin [53], Skowronski [63] and Stonier [64] to describe the dynamics of such games, with $F(u_1, u_2, x_0, t_0, t)$ a general control system for each admissible strategy $u_1 \in U_1$ of the first player and $u_2 \in U_2$ of the second player. This work is however unnecessarily restricted by the use of particular classes of admissible strategies and by the assumption of continuous dependence of F on these strategies. As in a control text, it would seem that greater generality and technical simplicity could be achieved if the strategies were not mentioned explicitly or if they were used only to index the corresponding dynamics. This latter idea and another concerning the nature of attainable points in a game context lead Kloeden [30] to describe game dynamics in terms of general control systems in quite a different way.

A lucid account of the subtle nuances in the concept of attainability in a game context is given by Roxin in [54]. The salient feature is the observation that if, say, the first player uses an admissible strategy $u_1 \in U_1$ then, without further knowledge of the other player's choice of strategy, he can determine the future evolution of the game no more accurately than within the attainability sets $F(u_1, U_2, x_0, t_0, t)$, to use the terminology from above. Similarly, if the second player chooses an admissible strategy $u_2 \in U_2$, then his knowledge of the future evolution of the game is no better than that it lies somewhere within the attainability sets $F(U_1, u_2, x_0, t_0, t)$. This suggests that the dynamical systems corresponding to these attainability sets should be used as the basic entities in a model of game dynamics. Each player could be assigned a set of general control systems $\mathcal{D}_i \subset F$, where, say, a general control system $F^1_{u_1} \in \mathcal{D}_1$ corresponds to the dynamical system $F(u_1, U_2, x_0, t_0, t)$ above. A further advantage of this idea is that the strategies u_i need only appear as indices for the sets of admissible dynamics $\mathcal{D}_i$, with any required topological conditions being imposed directly on these sets; for example, they could be assumed to be compact subsets of F for some convenient topology on F. To participate in such a game the first player chooses an admissible general control system $F^1_{u_1} \in \mathcal{D}_1$ and the second player an admissible general control system $F^2_{u_2} \in \mathcal{D}_2$. The actual dynamics of the game can then be reclaimed by taking the "intersection" of the general control systems $F^1_{u_1}$ and $F^2_{u_2}$, the existence of which would require additional conditions to be imposed on the sets $\mathcal{D}_1$ and $\mathcal{D}_2$ of admissible dynamics. At first it might seem that this intersection system $F_{u_1 u_2}$ should be defined as the (non-empty by assumption) intersection of attainability sets; that is,

$$F_{u_1 u_2}(x_0, t_0, t) = F^1_{u_1}(x_0, t_0, t) \cap F^2_{u_2}(x_0, t_0, t) .$$

This is, however, unsatisfactory because the dynamical system so defined need not satisfy the semigroup property of a general control system, a situation arising from the fact that an attainable point in the intersection of attainability sets not be attained by a trajectory lying entirely within these intersections. See Example 6.1 of [30]. This anomaly can however be avoided if instead the intersection is defined in terms of the trajectories common to both $F^1_{u_1}$ and $F^2_{u_2}$; that is,

$$F_{u_1 u_2}(x_0, t_0, t) = \cup \left\{ x = \phi(t); \phi \in \bigcap_{i=1}^{2} \Phi\left(x_0, t_0; F^i_{u_i}\right) \right\} .$$

The intersection $F_{u_1 u_2}$, which could appropriately be called the *trajectory-intersection* of $F^1_{u_1}$ and $F^2_{u_2}$, is then a general control system (except possibly for the backwards extendability axiom, which is unimportant: see Theorem 6.1 and Example 6.2 of [30]).

The above ideas are used in [30] to construct an axiomatic model for the dynamics of N-person games. This model has many advantages in the generality and simplicity it offers, but it does have a major shortcoming when applied to games, such as differential games, in which the dynamics are first given in terms of sets of strategies and a dynamics generating mapping. In such games it is possible for a trajectory to correspond to both a strategy pair (u_1, v_2) and a strategy pair (v_1, u_2). This trajectory would then be a trajectory of both $F^1_{u_1}$ and $F^2_{u_2}$, and hence of their trajectory-intersection $F_{u_1 u_2}$ even though it need not be generated by the strategy pair (u_1, u_2). An example of this involving a very simple differential game is given in the appendix of [30]. It emphasizes the need to explicitly retain the relationship between trajectories and their generating strategy pairs when describing game dynamics.

Finally, game dynamics have also been formulated using general control systems, albeit H-continuous in all variables, by Malafeev [34, 35] in the Soviet Union.

References

[1] Richard F. Arens, "A topology for spaces of transformations", *Ann. of Math.* (2) 47 (1946), 480-495.

[2] Е.А. Барбашин [E.A. Barbašin], "On the theory of general dynamical systems", *Učen. Zap. Moskov. Gos. Univ.* 135, *Mat. II* (1948), 110-133 (Russian).

[3] Е.А. Барбашин [E.A. Barbašin], "Dispersive dynamical systems", *Uspehi Mat. Nauk (N.S.)* 5 No. 4 (38) (1950), 138-139 (Russian).

[4] D.P. Bertsekas and I.B. Rhodes, "On the minimax reachability of target sets and target tubes", *Automatica* 7 (1971), 233-247.

[5] N.P. Bhatia, O. Hajek, *Local Semi-Dynamical Systems* (Lecture Notes in Mathematics, 90. Springer-Verlag, Berlin, Heidelberg, New York, 1969).

[6] N.P. Bhatia, G.P. Szegö, *Dynamical Systems: Stability Theory and Applications* (Lecture Notes in Mathematics, 35. Springer-Verlag, Berlin, Heidelberg, New York, 1967).

[7] F.S. de Blasi and J. Schinas, "Stability of multivalued discrete dynamical systems", *J. Differential Equations* 14 (1973), 245-262.

[8] T.F. Bridgland Jr., "Contributions to the theory of generalized differential equations. I", *Math. Systems Theory* 3 (1969), 17-50.

[9] T.F. Bridgland Jr., "Contributions to the theory of generalized differential equations. II", *Math. Systems Theory* 3 (1969), 156-165.

[10] И.У. Бронштейн [I.U. Bronšteĭn], "О динамических системах без единственности как полугруппах неоднозначных отобржений топологического пространства" [On dynamical systems without uniqueness, as semigroups of non-singlevalued mappings of a topological space], *Dokl. Akad. Nauk SSSR* 144 (1962), 954-957; *Soviet Math. Dokl.* 3 (1962), 824-827.

[11] И.У. Бронштейн, Б.А. Щербаков [I.U. Bronšteĭn, B.A. Ščerbakov], "Некоторые свойства устойчивых по Лагранжу воронок обобщнных динамических систем" [Certain properties of Lagrange stable funnels of generalized dynamical systems], *Izv. Akad. Nauk Moldav. SSSR* 5 (1962), 99-102.

[12] Б.М. Будак [B.M. Budak], "Dispersive dynamical systems", *Vestnik Moskov. Univ.* (1947), No. 8, 135-137 (Russian).

[13] Б.М. Будак [B.M. Budak], "The concept of motion in a generalized dynamical system", *Moskov. Gos. Univ. Uč. Zap.* 155, *Mat.* 5 (1952), 174-194 (Russian).

[14] D. Bushaw, "Dynamical polysystems and optimization", *Contributions to Differential Equations* 2 (1963), 351-365.

[15] D. Bushaw, "Dynamical polysystems: a survey", *Proc. US-Japan Seminar on Differential and Functional Equations*, 13-26 (Minneapolis, Minnesota, 1967. Benjamin, New York, 1967).

[16] M. Fukuhara, "Sur les familles de fonctions à une variable réelle", *J. Fac. Sci. Hokkaido Univ.* (I) 1 (1932), 163-209.

[17] Walter Helbig Gottschalk and Gustav Arnold Hedlund, *Topological Dynamics* (Amer. Math. Soc. Colloquium Publications, **36**. Amer. Math. Soc., Providence, Rhode Island, 1955).

[18] Hubert Halkin, "Topological aspects of optimal control of dynamical polysystems", *Contributions to Differential Equations* 3 (1964), 377-385.

[19] М.С. Изман [M.S. Izman], "Устойчивость и множества притяжения в лисперсных динамических системах" [Stability and sets of attraction in dispersive dynamical systems], *Mat. Issled.* 3 (1968), Vyp 3 (**9**), 60-78.

[20] М.С. Изман [M.S. Izman], "Устойчивость множеств и аттракторы в дисперсных динамических системах" [Stability of sets, and attractors in dispersive dynamical systems], *Mat. Issled.* 3 (1968), Vyp 4 (**10**), 51-77.

[21] М.С. Изман [M.S. Izman], "Применение второго метода Ляпунова для исследования устойчивости и асимптотической устойчивости множеств в дисперсных динамических системах" [An application of Ljapunov's second method for the investigation of stability and asymptotic stability of sets in dispersive dynamical systems], *Differencial'nye Uravnenija* 5 (1969), 1207-1217; *Differential Equations* 5 (1969), 883-891 (1972).

[22] М.С. Изман [M.S. Izman], "Об асимптотической устойчивости множеств в дисперсных динамических системах" [On the asymptotic stability of sets in dispersive dynamical systems], *Differencial'nye Uravnenija* 7 (1971), 615-621.

[23] John L. Kelley, *General Topology* (The University Series in Higher Mathematics. Van Nostrand, Toronto, New York, London, 1955).

[24] Peter Eris Kloeden, "On general semi dynamical systems" (PhD thesis, University of Queensland, Brisbane, 1974); see also: Abstract, *Bull. Austral. Math. Soc.* 13 (1975), 152-154.

[25] Peter Kloeden, "General control systems without backward extension", *Differential Games and Control Theory*, 49-58 (Lecture Notes in Pure and Applied Mathematics, **10**. Marcel Dekker, New York, 1974).

[26] P.E. Kloeden, "Generalised Markovian control systems", *J. Austral. Math. Soc.* 18 (1974), 485-491.

[27] P.E. Kloeden, "Asymptotic invariance and limit sets of general control systems", *J. Differential Equations* 19 (1975), 91-105.

[28] P.E. Kloeden, "Eventual stability in general control systems", *J. Differential Equations* 19 (1975), 106-124.

[29] Peter E. Kloeden, "Some remarks on relative stability", *J. Austral. Math. Soc. Ser. B* 19 (1975), 112-115.

[30] Peter E. Kloeden, "An attainability set model for dynamical games", *J. Differential Equations* 21 (1976), 197-230.

[31] P.E. Kloeden, "Topologies for general semi-dynamical systems", *J. Differential Equations* (to appear).

[32] H. Kneser, "Über die Lösungen eines Systems gewöhnlicher Differentialgleichungen, das der Lipschitzschen Bedingung nicht genügt", *S.-B. K. Preuss. Akad. Wiss. Berlin, Phys.-Math. Kl.* **1923**, 171-174.

[33] G.S. Ladde and S.G. Deo, "General dynamical systems and functional differential equations", *J. Mathematical and Physical Sci.* **3** (1969), 257-268.

[34] О.А. Малафеев [O.A. Malafeev], "О динамических играх с зависимыми движениями" [On dynamic games with dependent motions], *Dokl. Akad. Nauk SSSR* **213** (1973), 783-786; *Soviet Math. Dokl.* **14** (1973), 1792-1796 (1974).

[35] О.А. Малафеев [O.A. Malafeev], "Ситуацни равновесия в динамических играх" [Equilibrium situations in dynamic games], *Kibernetika* **1974**, No. 3 (1974), 111-118.

[36] L. Markus and E.B. Lee, *Foundations of Optimal Control Theory* (John Wiley & Sons, New York, London, Sydney, 1967).

[37] М.И. Минкевич [M.I. Minkevič], "Теория интепральных воронок в обобщнных динамических системах без предположения единственности" [The theory of integral funnels in generalized dynamical systems without a hypothesis of uniqueness], *Dokl. Akad. Nauk SSSR (N.S.)* **59** (1948), 1049-1052.

[38] М.И. Минкевич [M.I. Minkevič], "Theory of integral funnels in dynamical systems without uniqueness", *Uče. Zap. Moskov. Gos. Univ.* **135** Mat. II (1948), 134-151 (Russian).

[39] М.И. Минкевич [M.I. Minkevič], "Closed integral funnels in generalized dynamical systems without a hypothesis of uniqueness", *Uče. Zap. Moskov. Gos. Univ.* **163** Mat. **VI** (1952), 73-88 (Russian).

[40] Jozef Nagy, "Stability of sets with respect to abstract processes", *Mathematical Systems Theory and Economics,* 11 (Proc. Internat. Summer School, Varenna, 1967, 355-378. Lecture Notes in Operations Research and Mathematical Economics, **12**. Springer-Verlag, Berlin, Heidelberg, New York, 1969).

[41] В.В. Немыцкий, В.В. Степанов [V.V. Nemytskii and V.V. Stepanov], *Качественная Теория Дифферинциальных Уравнений* [*Qualitive Theory of Differential Equations*] (OGI2, Moscow, 1949. Transl. Princeton Mathematical Series, **22**. Princeton University Press, Princeton, New Jersey, 1960).

[42] B.G. Pachpatte, "Strict stability in general dynamical systems", *J. Differential Equations* **11** (1972), 464-473.

[43] T. Parthasarathy, *Selection Theorems and Their Applications* (Lecture Notes in Mathematics, **263**. Springer-Verlag, Berlin, Heidelberg, New York, 1972).

[44] В.И. Пономарев [V.I. Ponomarev], "Новое пространство замкнутых множеств и многозначные непрерывные отображения бикомпантов" [A new space of closed sets and multivalued continuous mappings of bicompacta], *Mat. Sb. (N.S.)* **48 (90)** (1959), 191-212; *Amer. Math. Soc. Transl.* (2) **38** (1964), 95-118.

[45] E.O. Roxin, "Axiomatic foundation of the theory of control systems", *Proc. Second Internat. Conf. IFAC*, Basel 1963, 640-644 (Butterworths, London, 1964).

[46] E.O. Roxin, "Dynamical systems with inputs", *Proc. Symposium on System Theory*, New York 1965, 105-111 (Microwave Research Institute Symposia Series, **15**. Polytechnic Press, Polytechnic Institute of Brooklyn, Brooklyn, New York (distributed by Interscience [John Wiley & Sons], New York, London), 1965).

[47] Emilio Roxin, "Stability in general control systems", *J. Differential Equations* **1** (1965), 115-150.

[48] Emilio Roxin, "On generalized dynamical systems defined by contingent equations", *J. Differential Equations* **1** (1965), 188-205.

[49] Emilio Roxin, "On stability in control systems", *SIAM J. Control* **3** (1965), 357-372 (1966).

[50] Emilio Roxin, "Local definition of generalized control systems", *Michigan Math. J.* **13** (1966), 91-96.

[51] Emilio Roxin, "On asymptotic stability in control systems", *Rend. Circ. Mat. Palmero* (2) **15** (1966), 193-208.

[52] Emilio Roxin, "On finite stability in control systems", *Rend. Circ. Mat. Palmero* (2) **15** (1966), 273-282.

[53] Emilio Roxin, "Axiomatic approach in differential games", *J. Optimization Theory Appl.* **3** (1969), 153-163.

[54] Emilio O. Roxin, "Some global problems in differential games", *Global Differential Dynamics*, 103-116 (Lecture Notes in Mathematics, **235**. Springer-Verlag, Berlin, Heidelberg, New York, 1971).

[55] C. Ryll-Nardzewski, "A theory of pursuit and evasion", *Advances in Game Theory* (Annals of Mathematics Studies, **52**, 113-126. Princeton University Press, Princeton, New Jersey, 1964).

[56] S.H. Saperstone, "On the attainable set of conditional probability measures in stochastic control problems", *Internat. J. Systems Sci.* **7** (1976), 131-135.

[57] Peter Seibert, "Stability under perturbations in generalized dynamical systems", *Internat. Sympos. Nonlinear Differential Equations and Nonlinear Mechanics*, 463-473 (Academic Press, New York, 1963).

[58] George R. Sell, "On the fundamental theory of ordinary differential equations", *J. Differential Equations* **1** (1965), 370-392.

[59] К.С. Сибирский [K.S. Sibirskii], *Введение в Топологическую Динамику* (Redakcionno-Izdat. Otdel Akad. Nauk Moldav. SSR, Kishinev, 1970). English Transl: *Introduction to Topological Dynamics* (translated by Leo F. Boron. Noordhoff, Leyden, 1975).

[60] К.С. Сибирский, Н.К. Чебан [K.S. Sibirskiĭ and N.K. Čeban], "Классификация устойчивых по.пуассону точек в диоперсных динамических системах" [Classification of Poisson stable points in dispersive dynamical systems], *Dokl. Akad. Nauk SSSR* **209** (1973), 801-804.

[61] К.С. Сибирский, И.А. Чиркова [K.S. Sibirskiĭ and I.A. Čirkova], "Discontinuous dispersive dynamical systems", *Mat. Issled.* **6** (1971), Uyp. 4 (**22**), 165-175 (Russian).

[62] К.С. Сибирский, И.А. Чиркова [K.S. Sibirskiĭ and I.A. Čirkova], "К теории разрывных дисперсных динамических систем" [On the theory of discontinuous dispersive dynamical systems", *Dokl. Akad. Nauk SSSR* **211** (1973), 1302-1305; *Soviet Math. Dokl.* **14** (1973), 1256-1261 (1974).

[63] J.M. Skowronski, "Qualitative hazard avoiding Lyapunov games", Preprint, **42**. University of Queensland, Brisbane, 1973.

[64] R.J. Stonier, "Ultimate boundedness of a two-person game using Lyapunov theory", Preprint, **41**. University of Queensland, Brisbane, 1973.

[65] G.P. Szegö and G. Treccani, "Flow without uniqueness near a compact strongly invariant set", *Boll. Un. Mat. Ital.* (4) **2** (1969), 113-124.

[66] G.P. Szegö, G. Treccani, *Semigruppi di Trasformazioni Multivoche* (Lecture Notes in Mathematics, **101**. Springer-Verlag, Berlin, Heidelberg, New York, 1969).

[67] G.P. Szegö and G. Treccani, "An abstract formulation of minimization algorithms", *Differential Games and Related Topics*, 279-297 (North-Holland, Amsterdam, London, 1971).

[68] P.P. Varaiya, "On the existence of solutions to a differential game", *SIAM J. Control* **9** (1967), 153-162.

[69] Taro Yoshizawa, *Stability Theory by Liapunov's Second Method* (Publications of the Mathematical Society of Japan, **9**. The Mathematical Society of Japan, Tokyo, 1966).

[70] S.K. Zaremba, "Sur certaines familles de courbes en relation avec la théorie des équations différentielles", *Ann. Soc. Polon. Math.* **15** (1937), 83-100.

[71] В.И. Зубов [V.I. Zubov], *Методы А.М. Ляпунова и их Применение* (Izdat. Leningrad Univ., Moscow, 1957). English Transl: *Methods of A.M. Lyapunov and Their Application* (Translated by Leo F. Boron. Noordhoff, Groningen, 1964).

THE BANG-BANG PRINCIPLE

Igor Kluvánek and Greg Knowles

The aim of this survey is to present a framework in which statements that a control leading to a desired effect takes its values from extreme points of the admissible set can be expressed in a concise and unified manner. We emphasize results concerning systems with infinite-dimensional space of states. Almost no attention is paid to the history of the subject. The list of references is by no means exhaustive or even indicative about the literature concerning the subject. The inclusion or otherwise of an item in no way represents our reflection on its importance. Possibly, by following the references in the quoted works, a better picture about the history of the subject can be obtained. A more serious omission, dictated by the limitations of the space and time, is the fact that deeper applications are not discussed. For example, the description of effective methods for numerical calculation of optimal controls based on their taking on as values the extreme points of the admissible set would be impracticable in an article of this scope. However, we wish to stress that it is mainly because of the nontraditional applications that the theme of this survey is so interesting and attractive.

1. Control Systems

Let X be a real locally convex topological vector space. The space X will serve as the space of possible states of considered systems. It will always be assumed quasi-complete.

If T is a set and S a σ-algebra of subsets of T, we denote by $M(S)$ or, shortly, by M the set of all S-measurable functions on T. By $BM(S)$ or by BM is denoted the set of all bounded S-measurable functions on T.

An X-valued vector measure is a σ-additive map $m : S \to X$ whose domain S is a σ-algebra of subsets of a set T. The set $m(S) = \{m(E) : E \in S\}$ is its range. A set $E \in S$ is termed m-null if $m(F) = 0$ for every $F \in S$, $F \subset E$. A function f (even vector-valued) is termed m-null if the set $\{t : f(t) \neq 0\}$ is m-null. The integral of a function f with respect to the vector measure m is denoted, for short, by $m(f)$.

Given $E \in S$, we denote $S_E = \{F : F \in S, F \subset E\}$. Further, m_E is the restriction of m to S_E, that is, $m_E(F) = m(F)$, for every $F \in S_E$.

A net $\{E_\alpha\}$ of sets in S is said to be m-convergent to an element E of S (respectively, to be m-Cauchy) if, for every neighbourhood U of 0 in X, there is an index α_U such that $m(F) \in U$, for every set $F \subset E_\alpha \Delta E$ (respectively, $F \subset E_\alpha \Delta E_\beta$), $F \in S$, whenever $\alpha_U \leq \alpha$ (respectively, $\alpha_U \leq \alpha$, $\alpha_U \leq \beta$). The vector measure $m : S \to X$ is said to be closed if S is m-complete, that is, if every m-Cauchy net of sets in S is m-convergent to a member of S.

Sequences $f = (f_i)$ of S-measurable (real-valued) functions are termed controls. A control f is interpreted as determining the values $f_i(t)$, $i = 1, 2, \ldots$, of a sequence of controlling parameters at any $t \in T$. In other words, f is the law of a control policy.

A sequence $m = (m_i)$ of closed vector measures $m_i : S \to X$, $i = 1, 2, \ldots$, will be called a control system or, more precisely, a linear control system, if the sum $\sum_{i=1}^{\infty} x_i$ is convergent for any $x_i \in m_i(S)$, $i = 1, 2, \ldots$. Because $0 \in m_i(S)$, $i = 1, 2, \ldots$, this convergence is unconditional. We denote $m_E = ((m_i)_E)$, for any $E \in S$.

The effect of a control $f = (f_i)$ on a system $m = (m_i)$ or, alternatively, the response of the system m to the control f is the point

(1) $$m(f) = \sum_{i=1}^{\infty} \int_T f_i(t) dm_i(t) = \sum_{i=1}^{\infty} m_i(f_i)$$

of the state-space X, provided it exists. It does exist in a large class of cases, for example if the control f is uniformly bounded, meaning that $|f_i(t)| \leq k$, for some k and every $i = 1, 2, \ldots$ and every $t \in T$.

This scheme is sufficiently general to include, as special cases, the majority of control processes in which the response of the controlled system depends linearly on the control policy. The value $m_j(E)$ of the jth measure at a set $E \in S$ is the response of the system to the control $f = (f_i)$, where f_j is the characteristic function of the set E and f_i are zero-functions for every $i \neq j$. The linearity (the principle of superposition) gives that $m_j(E)$ depends additively on E. The assumption of σ-additivity is the result of continuity in a sense of the dependence of the response on the control. The linearity and a sort of continuity implies then also (1).

The assumption that the vector measures be closed does not represent any serious restriction. If the space X is metrizable the assumption is automatically fulfilled. Also, if the vector measure is an indefinite (Pettis) integral of vector function with respect to a positive measure, the case most frequently appearing in applications, then the vector measure is closed. On the other hand, the integration with respect to a closed vector measure leads to a complete space of integrable functions analogously to the L^1-space with respect to a positive measure.

In the final analysis, the assumptions that the vector measures m_i, $i = 1, 2, \ldots$, constituting a control system $m = (m_i)$ be σ-additive and closed are warranted by two facts, namely

(i) in cases of interest these assumptions are fulfilled, and

(ii) the assumptions are the basis of a theory rich with useful results not possible without them.

A more detailed discussion of all these concepts can be found in [6].

Systems controlled by a finite number, or perhaps, by only one parameter, fall into the introduced scheme. They may be described by control systems $m = (m_i)$ such that all but only a finite number of vector measures m_i are identically zero.

Given a control system $m = (m_i)$, a control $f = (f_i)$ will be said m-null if f_i is m_i-null function for every $i = 1, 2, \ldots$. Controls f and g are m-equivalent if $f - g$ is m-null. A control f is said to have m-essentially a property if there exists a control g with the property which is m-equivalent to f. We say that there is a m-essentially unique control with a property if all controls with the property are m-equivalent to each other.

2. Attainable Set

Usually there are some restrictions imposed on the controls. Perhaps all situations encountered can be described as follows.

We denote by $\mathbb{R}^\infty$ the Cartesian product of a sequence (indexed by $i = 1, 2, \ldots$) of copies of the real-line. It may be treated as a topological vector space under the product topology.

If F is a function on T whose values are subsets of $\mathbb{R}$, the real-line, the set of all S-measurable functions f on T such that $f(t) \in F(t)$, for every $t \in T$, is denoted by $M_F(S)$ or, shortly, by M_F. Similarly, if F is a function on T whose values are subsets of $\mathbb{R}^\infty$, the set of all controls $f : T \to \mathbb{R}^\infty$ such that $f(t) \in F(t)$, for every $t \in T$, is denoted by $M_F(\mathbb{R}^\infty, S)$ or, shortly, by $M_F(\mathbb{R}^\infty)$ or even by M_F.

Given a function F on T whose values are subsets of $\mathbb{R}^\infty$, the elements of M_F are termed admissible controls. If a control system m is given then the set

$$A_F(m) = \{m(f) : f \in M_F, m(f) \text{ exists}\}$$

is the attainable set of the control system m subject to the restriction F.

A function F defined on T whose values are subsets of $\mathbb{R}^\infty$ will be called bounded if there exists a compact set $K \subset \mathbb{R}^\infty$ such that $F(t) \subset K$, for every $t \in T$.

Denote by $cc\,\mathbb{R}^\infty$ the family of all compact convex subsets of $\mathbb{R}^\infty$. A set-valued function $F : T \to cc\,\mathbb{R}^\infty$ is called measurable if, for every sequence (β_i), with all terms but a finite number of them vanishing, the function

$$t \mapsto \sup\left\{\sum_{i=1}^\infty \beta_i \alpha_i : (\alpha_i) \in F(t)\right\}, \quad t \in T,$$

is S-measurable.

The symbol $\mathrm{ex}\,W$ denotes the set of all extreme points of a subset W of a vector space. For a set-valued function $F : T \to cc\,\mathbb{R}^\infty$, the set-valued function $\mathrm{ex}\,F$ is defined by $(\mathrm{ex}\,F)(t) = \mathrm{ex}\,F(t)$, $t \in T$.

We denote by I the Cartesian product of intervals $I_i = [-1, 1]$, $i = 1, 2, \ldots$. At the same time, I will stand for the constant set-valued function $F : T \to cc\,\mathbb{R}^\infty$ such that $F(t) = I$, for $t \in T$. This set-valued function represents a classical constraint resulting in bounded amplitude controls.

THEOREM 1. *Let $m = (m_i)$ be a control system and $F : T \to cc\,\mathbb{R}^\infty$ a bounded measurable set-valued function.*

Then $A_F(m)$ is a convex weakly compact subset of X.

If x is an extreme point of $A_F(m)$ then there is an m-essentially unique

$f \in M_F$ for which $x = m(f)$. Moreover, $m_E(f)$ is an extreme point of $A_F(m_E)$, for every $E \in S$.

The proof of this theorem can be found in [6, Theorems IX.1.1 and IX.2.1].

From the point of view of this survey, it would be desirable if from $x \in \text{ex } A_F(m)$ and $x = m(f)$ with $f \in M_F$ we could deduce that f belongs m-essentially to $M_{\text{ex}F}$. (An erroneous statement to this effect is contained in Theorem IX.2.1 in [6].) For a large class of set-valued functions F it is indeed true that if x is an extreme point of $A_F(m)$ and if $x = m(f)$ with $f \in M_F$, then F belongs m-essentially to $M_{\text{ex}F}$. This class includes, in particular, measurable and bounded set-valued functions $F : T \to CC\,R^\infty$ such that the boundary of any value $F(t)$ does not contain an 'oblique' segment. This covers, among others, the case of the bounded amplitude controls, that is, $F = I$. More precise formulation and the proof can be found in [8]. A complete clarification of the situation when the extreme points of the attainable sets are the results of the application of controls belonging to $M_{\text{ex}F}$ would be desirable.

The requirement of boundedness of F in Theorem 1 and related considerations can be relaxed somewhat; see [8].

3. Liapunov Systems

A vector measure $m : S \to X$ is called a Liapunov vector measure if the set $m(S_E)$ is convex and weakly compact for each $E \in S$.

THEOREM 2. *Let* $m : S \to X$ *be a closed vector measure. Each of the following conditions is both necessary and sufficient for* m *to be Liapunov:*

(i) *for every set* $E \in S$ *which is not* m-null, *there exists a bounded* S-measurable function f which is not m-null on E and $m_E(f) = 0$;

(ii) *for every bounded* S-measurable function g which is not m-null, there exists a bounded S-measurable function f such that fg is not m-null but $m(fg) = 0$.

This theorem represents a quite effective criterion of Liapunovness for vector measures (systems with one independent control parameter). It was proved, in this generality, in [7]. The proof is reproduced in [6] together with some comments concerning its predecessors. Many initial value or boundary value control problems are amenable to this criterion [6, Section V.7].

To illustrate the situation, let us consider the value $x(t) = u(0, t)$, at every t in a time-interval J, of the solution u of the problem

$$\frac{\partial^2 u}{\partial t^2} = c^2 \Delta u \; ; \quad u(p, 0) = 0 \; , \quad \frac{\partial u}{\partial t}(p, 0) = f(p) \; , \quad p \in \mathbb{R}^3 \; ,$$

where $c > 0$ is a given number, Δ denotes the Laplacian with respect to the space coordinates in $\mathbb{R}^3$, and f is a function on $\mathbb{R}^3$ interpreted as a control. The formula

$$x(t) = \frac{t}{4\pi} \int_0^{2\pi} \int_0^{\pi} f(ct, \theta, \lambda) \sin\theta \, d\theta \, d\lambda \; ,$$

holding for almost every $t \geq 0$ (assuming that points in $\mathbb{R}^3$ are given by their spherical coordinates) permits the use of Theorem 2 to show that, if J is bounded, the relation between f and x is given by the integration with respect to a Liapunov $L^1(J)$-valued vector measure.

An example of a boundary value control is furnished by the problem of finding the steady state temperature distribution in the half-space $M = \{(\xi, \eta, \zeta) \in \mathbb{R}^3 : \zeta \geq 0\}$ controlled in the plane $\{(\xi, \eta, \zeta) : \zeta = 0\}$ to be $f(\xi, \eta)$, $(\xi, \eta) \in T = (-\infty, \infty) \times (-\infty, \infty)$. The temperature u satisfies the boundary value problem

$$\Delta u = 0 \quad \text{in } M \; ; \quad u(\xi, \eta, 0) = f(\xi, \eta) \; , \quad (\xi, \eta) \in T \; .$$

We wish to determine the temperature in the plane T so as to obtain a desired temperature $x(\zeta)$ at points $(0, 0, \zeta)$, $\zeta > 0$, inside M. Then

$$x(\zeta) = \frac{\zeta}{2\pi} \int_T \frac{f(\xi,\eta) \, d\xi \, d\eta}{(\zeta^2 + \xi^2 + \eta^2)^{3/2}} \; ,$$

for every $\zeta > 0$. Theorem 2 implies that x, considered as an element of the space $C_0([0, \infty))$ of continuous functions on $[0, \infty)$ vanishing at ∞, is given as the integral of f with respect to a Liapunov $C_0([0, \infty))$-valued vector measure.

Given a set-valued function $F : T \to CC\,\mathbb{R}^\infty$, a control system m is called F-Liapunov if $A_F(m_E) \in A_{exF}(m_E)$, for every $E \in S$.

THEOREM 3. *A control system* $m = (m_i)$ *is I-Liapunov if and only if every vector measure* m_i, $i = 1, 2, \ldots$, *is Liapunov.*

If the control system m is F-Liapunov then the only points $x \in A_F(m)$ with an m-essentially unique $f \in M_F$ for which $x = m(f)$ are the extreme points of $A_F(m)$.

If X is a space such that, for every control system (n_i) consisting of

Liapunov vector measures, the vector measure $\sum_{i=1}^{\infty} n_i$ *is Liapunov, then any control system* $m = (m_i)$ *formed by X-valued Liapunov vector measures is F-Liapunov for each bounded measurable set-valued function* $F : T \to CC\mathbb{R}^{\infty}$.

The proof of the first statement of this theorem is in [5]. The other statements can be found in [6, Sections IX.2 and IX.3].

The last statement of this theorem applies to finite-dimensional spaces. It is easy to exhibit spaces which do not satisfy this statement. But the question whether the sum of a *finite* number of Liapunov vector measures is Liapunov seems to be still unanswered.

4. A Necessary Condition for Optimality

In the case of an F-Liapunov control system every point of the attainable set $A_F(m)$ is the result of the application of a control f such that $f(t) \in \text{ex } F(t)$, $t \in T$. If the system m is not F-Liapunov then not all points of $A_F(m)$ can be so obtained but points $x \in A_F(m)$ of prime interest could still be obtained by the use of controls belonging to $M_{\text{ex}F}$. The points of $A_F(m)$ which are optimal in some sense belong usually to the boundary of $A_F(m)$, or even to some distinguished part of it, and they are likely to correspond to controls from $M_{\text{ex}F}$.

Let X' be the continuous dual space of X. If a functional $x' \in X'$ achieves its maximal value on a set $A \subset X$ at a point $x \in A$, then x' is said to support A at x, also x is called a support point of A and the hyperplane $\{y : \langle x', y \rangle = \langle x', x \rangle, y \in X\}$ is called a supporting hyperplane of A.

A point $x \in A$ is called an exposed point of A if there is a functional $x' \in X'$ such that $\langle x', y \rangle < \langle x', x \rangle$ for every $y \in A$, $y \neq x$. The functional x' is then said to expose A at x. The set of exposed points of A is denoted by $\exp A$.

Given a vector measure $m : S \to X$ and a functional $x' \in X'$, by $\langle x', m \rangle$ is denoted the real-valued measure on S defined by $\langle x', m \rangle(E) = \langle x', m(E) \rangle$, for every $E \in S$. If $m = (m_i)$ is a control system then we put $\langle x', m \rangle = (\langle x', m_i \rangle)$.

Let $m = (m_i)$ be a control system and let $F : T \to CC\mathbb{R}^{\infty}$ be a bounded measurable set-valued function. If the functional $x' \in X'$, $x' \neq 0$, supports $A_F(m)$ at a point $x = m(f)$ with $f \in M_F$, then

(2) $\qquad \langle x', m(f) \rangle = \max\{\langle x', m(g) \rangle : g \in M_F\}$

there is $\langle x', m\rangle$-essentially unique control $f \in M_F$ such that $x = m(f)$ and this control belongs $\langle x', m\rangle$-essentially to M_{exF}.

Some classical problems lead to condition (2). For example, if X is a Banach space and y is a point not belonging to $A_F(m)$, then the problem of finding $f \in M_F$ such that $\|m(f)-y\| \leq \|x-y\|$, for every $x \in A_F(m)$, leads to (2). In this case, $m(f)$ is a support point of $A_F(m)$. This type of problem was considered already by Egorov [2] in connection with problems of the conduction of heat and by other authors. (The problem was mentioned also at this conference by S. Gustafson.) Recently in a more general setting it was taken up in [10]. It is obvious that this type of problem is susceptible to generalizations. Instead of distance from a point some other functions could be minimized.

Also time-optimal problems lead to (2). In that case, we have a control system depending on a parameter interpreted as time and the problem is to reach a convex set W in a minimum time. If W has non-empty interior then the point of first contact with W is a support point of the attainable set. In some cases, such as if $\dim X < \infty$, the set W can be allowed to reduce to a single point. In general one has to require that W has an interior point. It was pointed out by Egorov that, especially in infinite-dimensional problems, it is more realistic to try to approximate the desired point with a prescribed accuracy rather than to reach it exactly; it may not even be possible to reach the desired point at all. A class of problems in this vein was considered in [10]. It may be of interest to note examples in Section 7 to follow.

5. Exposable Systems

The condition (2) determines f only up to $\langle x', m\rangle$-equivalence. It can happen that controls f_1 and f_2 which are not m-equivalent both satisfy it. Now we shall consider situations when (2) suffices to determine m-essentially the control f.

The absolute continuity of a real- or vector-valued measure m with respect to a real- or vector-valued measure n means that every n-null set is m-null.

A vector measure $m : S \to X$ is said to be exposable in X if it is absolutely continuous with respect to every measure $\langle x', m\rangle$ with $x' \in X'$, $x' \neq 0$.

The term "exposable" is suggested by a result of Anantharaman [1] saying that a functional $x' \in X'$ exposes the range $m(S)$ of a vector measure $m : S \to X$ if and only if m is absolutely continuous with respect to $\langle x', m\rangle$. Hence the vector measure m is exposable in X if and only if every non-zero functional $x' \in X'$ exposes its range $m(S)$.

Assume that the vector measure $m : S \to X$ is absolutely continuous with respect to a non-negative measure μ on S. If the measure μ is absolutely continuous with respect to $\langle x', m \rangle$, for every $x' \in X'$ such that $x' \neq 0$, then the vector measure m is said to be exposable in X with respect to μ.

The terminology just introduced differs from that used hitherto in the literature. In [11] systems represented by $\mathbb{R}^n$-valued measures exposable with respect to one-dimensional Lebesgue measure introduced. Such systems are called "normal" in [11] and also in [4] where they are extensively studied. Systems giving rise to exposable vector measures are called "essentially normal" in [9]. We try to avoid the over-used term "normal" with qualifications.

A vector measure $m : S \to X$ is said to be controllable in the space X if, for every $x \in X$, there exists a bounded S-measurable function f such that $m(f) = x$. This is to say, m is controllable in X if the set $\{m(f) : f \in BM(S)\}$ is equal to the whole of the space X.

A vector measure $m : S \to X$ is said to be approximately controllable in the space X if the closure of the set $\{m(f) : f \in BM(S)\}$ is equal to X. Otherwise expressed, the vector measure m is approximately controllable in X if, for every $x \in X$ and every neighbourhood U of 0 in X, there exists a bounded S-measurable function f such that $m(f) \in x + U$.

The concepts of exposable, controllable and approximately controllable measures all depend on the space X. It very well may happen that a vector measure which is not exposable or controllable or approximately controllable on X is such in a subspace of X ([10], [3]).

THEOREM 4. *A vector measure* $m : S \to X$ *is exposable in the space* X *if and only if, for every set* $E \in S$ *which is not* m-*null, the vector measure* m_E *is approximately controllable in* X.

The proof is an almost immediate consequence of the definitions and the Hahn-Banach theorem.

A control system $m = (m_i)$ will be termed exposable or exposable with respect to a measure μ if all vector measures m_i, $i = 1, 2, \ldots$, constituting it have the corresponding property.

In the following theorem $F : T \to CC\mathbb{R}^\infty$ will be a bounded measurable set-valued function such that if $x \in \text{ex } A_F(m)$ and $x = m(f)$, $f \in M_F$, then $f \in M_{\text{ex}F}$ (see the comments after Theorem 1 in Section 2).

THEOREM 5. *If* m *is an exposable control system then the control* $f \in M_F$ *satisfying (2), for some* $x' \in X'$, $x' \neq 0$, *is* m-*essentially unique,* m-*essentially belongs to* $M_{\text{ex}F}$, *and if* $m(f) = m(g)$, *for some* $g \in M_F$, *then* f *and* g *are* *m-equivalent.*

A control system m is exposable in X if and only if every non-zero functional $x' \in X'$ which supports $A_I(m)$ at a point exposes it at that point.

The proof of this theorem is essentially in [8].

6. Considerations of Analycity

Theorem 4 as it stands may not be sufficiently easy to use for showing that a system is exposable. While there are criteria available for proving that a system is approximately controllable (see [14]), the test represented by Theorem 4 requires the approximate controllability of the restriction of a system on every non-null set. This may be difficult to establish. If, however, the measure has an analytic density the situation becomes more amenable.

Assume that T is an analytic manifold and that μ is a regular positive measure on T. Let S be the σ-algebra of Borel sets on T and let $m : S \to X$ be a vector measure whose density with respect to μ is an analytic function $\varphi : T \to X$.

Under these assumptions, if m is approximately controllable in X then it is exposable in X with respect to μ.

In fact, if the set $E \in S$ is not μ-null and $\langle x', m(G) \rangle = 0$ for every $G \subset E$, $G \in S$, then $\langle x', \varphi(t) \rangle = 0$, for μ-almost every $t \in E$. By the analyticity of φ, it is $\langle x', \varphi(t) \rangle = 0$, for every $t \in T$. Hence $\langle x', m(G) \rangle = 0$ for every $G \in S$, and, by the approximate controllability of m, it must be $x' = 0$. Hence m_E is approximately controllable in X for every $E \in S$ which is not μ-null.

The situation of an analytic density occurs often in applications. The first instances of the use of exposability (equals normality) arguments (such as reported in [4], [9]) concern vector measures with analytic densities.

Consider the control problem whose state equation is

(3) $$\dot{y}(t) = A(t)y(t) + f(t)b(t) , \quad y(0) = y_0 ,$$

where $A(t)$ is a (in general not continuous) operator on X, $b(t)$ is a column of vectors in X, for every $t \geq 0$, and y_0 is a fixed element of X. Let $t \mapsto S(t)$, $t \in [0, \infty)$, be the principal solution of the corresponding homogeneous operator equation. That is, $S(t) : X \to X$, $t \geq 0$, is such that

$$\dot{S}(t) = A(t)S(t) , \quad t > 0 ; \quad \lim_{t \to 0+} S(t)x = x = S(0)x ,$$

for every $x \in X$. Let

$$y(t, f) = S(t)y_0 + \int_0^t S(t)\bigl(S(\tau)\bigr)^{-1} f(\tau)b(\tau)d\tau , \quad t > 0 .$$

The function $\bigl(S(\ )\bigr)^{-1}b(\ )$ may turn out to be analytic in $(0, t)$. The situation is well investigated if $A(t) = A$ is a constant operator. Then $t \mapsto S(t)$, $t \in [0, \infty)$, is a semi-group. Assuming that b is a single vector which is an analytic vector of the semi-group S, our arguments apply.

If the control f is confined to the interval $[-1, 1]$, say, and if $y(t, f)$ is a supporting point of the attainable set, then the control f is given as $f(\tau) = \text{sgn}\langle x', S(t-\tau)b\rangle$, $\tau \in [0, t]$. Also, f can change the sign only countably many times accumulating at t. If $t \mapsto S(t)b$ happens to have an analytic extension into negative values of t, then f can have only finite number of switches (changes of sign). This last situation arises when S is an analytic group (that is, A is a bounded operator), in particular in finite-dimensional spaces [4].

Cases when (3) represents an initial-boundary value problem of parabolic type were considered in [10].

It may be noted that the full force of analyticity is not used in these considerations. It may happen that the requirement of the existence of an analytic density can be relaxed and some other uniqueness results (concerning solutions of partial differential equations) can be used instead (see, for example, [8, Theorem 4]).

7. Examples

Control problems with distributed control (described by the operator equation (3); the forcing term is controlled) suggest in some way problems concerning systems with a finite number of degrees of freedom. If the assumptions described in Section 6 are satisfied solutions of such problems resemble in some aspects those of finite-dimensional systems as in [4]. The equation (3) could represent either hyperbolic or parabolic boundary and initial value problems.

The situation is different in the case when boundary values are controlled. It should be noted, in this connection, that a hyperbolic boundary value control problem can not fit into the scheme described in Sections 4 and 5. Indeed, as showed by Russell [12, 13], the corresponding control systems are never exposable. This is explained by Theorem 4 and the fact that certain time has to lapse (namely the time needed for the wave to travel across the considered body) till the system becomes controllable. Parabolic problems are more amenable. Let us consider the following control problem, a special case of one discussed in [9].

Let Ω be a bounded domain in $\mathbb{R}^l$. For a multi-index $\alpha = (\alpha_1, \ldots, \alpha_l)$ of non-negative integers we put $|\alpha| = \alpha_1 + \ldots + \alpha_l$ and $D^\alpha = D_1^{\alpha_1} \ldots D_l^{\alpha_l}$, where D_j is the partial differentiation with respect to the jth coordinate, $j = 1, 2, \ldots, l$. Let

$$u \mapsto Lu = \frac{\partial u}{\partial \tau} - \sum_{|\alpha| \leq 2} a_\alpha(\xi, \tau) D^\alpha u$$

be a differential operator parabolic in the sense of Petrovski in $\overline{\Omega} \times (0, \infty)$. Let $t > 0$; let the function g be given on $\Omega \times [0, t]$; let the function φ be given on Ω and let the function a be given on $T = \partial\Omega \times (0, t)$. Denote by $x(\xi) = u(\xi, t)$ the value at any $\xi \in \Omega$ of the solution u at t of the problem

$$Lu = g(\xi, \tau), \quad (\xi, \tau) \in \Omega \times (0, t);$$

$$u(\xi, 0) = \Omega(\xi), \quad \xi \in \Omega;$$

$$\frac{\partial u}{\partial n} + a(\xi, \tau)u = f(\xi, \tau), \quad (\xi, \tau) \in T,$$

where $\partial/\partial n$ is the outward transversal differentiation on the lateral boundary of $\Omega \times (0, t)$ and f is a bounded measurable function on T interpreted as control.

If $\partial\Omega$ is an analytic manifold and the coefficients a_α of the operator L and the function a are analytic, then the correspondence between f and x, considered as an element of the space $X = L^p(\Omega)$ ($1 \leq p < \infty$), is a control system (vector measure) which is exposable with respect to natural surface area measure on $T = \partial\Omega \times (0, t)$. The time-optimal problem or the problem of the best approximation of an element of the space X, mentioned both in Section 6, are easily tractable.

A different type of example is obtained if we interpret a transmission line (or a recording device) as a control system. Let us consider such a line disregarding all distortions except the fact that the line cuts off all frequencies above the level $\sigma > 0$. It means that the Fourier transform of the received (or recorded) signal is obtained from the transmitted signal by multiplication by the characteristic function of the interval $[-\sigma, \sigma]$.

Assume that the admissible inputs into the line are of duration t and are bounded in modulus by 1. That is, an input is a Borel function f in the interval $T = [0, t]$ with values in the interval $[-1, 1]$ (when its Fourier transform is considered then the value outside T is taken to be 0); S will be the family of Borel sets in T. We interpret inputs as controls. Accordingly, the outputs (or the records) corresponding to an input f will be denoted by $m(f)$. The outputs are interpreted as elements of the space X consisting of all functions x on $(-\infty, \infty)$ such that

$$\|x\| = \sup\left\{\left(\frac{1}{2\delta}\int_{\tau-\delta}^{\tau+\delta} x^2(\xi)d\xi\right)^{\frac{1}{2}} : \tau \in (-\infty, \infty)\right\} < \infty,$$

where δ is a fixed positive constant. It seems that it was Vituškin (see, for example, [15]), who pointed at the relevance of this space to this type of problems.

The desired or useful signal is a function x, an element of the space X, which vanishes outside the interval T. Hence, apart from the trivial exception, the useful signal can not belong to the attainable set $\{m(f) : f \in M_{[-1,1]}(S)\}$ which consists of entire functions only. (That a useful signal can not be an analytic function follows also from the observation that otherwise all information which it carries would be carried already by its restriction to an arbitrarily short sub-interval of T.)

The problem is now to find an admissible input (control) f such that the distance in the space X of the output $m(f)$ from the signal x be minimal possible.

It turns out that the system (vector measure) m is exposable. Consequently, the solution of our problem is an input f which takes only the values -1 and 1.

8. Remark on Non-linear Case

Many results mentioned earlier extend to control problems where the dynamics of the system are still linear but controls enter non-linearly. The case of one independent control f can be described by a vector measure $m : S \to X$ and a bounded Carathéodory function $h : T \times \mathbb{R}^1 \to \mathbb{R}^1$. The response of the system to control f is then

$$\int_T h(t, f(t)) dm(t) .$$

Let U be a fixed compact set $\subset \mathbb{R}^1$. The admissible controls are ones taking values in U. Let $F(t) = \text{co}\{h(t, u) : u \in U\}$, $t \in T$. If x is a support point of $A_F(m)$ with supporting functional x', then

$$x = \int_T h(t, f(t)) dm(t)$$

with f such that $h(t, f(t)) \in \text{ex } F(t)$, for $\langle x', m \rangle$-almost every $t \in T$ ([8], [10]).

References

[1] R. Anantharaman, "On exposed points of the range of a vector measure", *Vector and Operator Valued Measures and Applications* (Proc. Sympos. Snowbird Resort, Alta, Utah, 1972, 7-22. Academic Press [Harcourt Brace Jovanovich], New York and London, 1973).

[2] Ю.В. Егоров [Ju.V. Egorov], "О некоторых задачах теории оптимального управления" [Some problems in theory of optimal control], *Dokl. Akad. Nauk SSSR* **145** (1962), 720-723.

[3] Hubert Halkin, "A generalization of LaSalle's 'bang-bang' principle", *SIAM J. Control* **2** (1964), 199-202.

[4] Henry Hermes and Joseph P. LaSalle, *Functional Analysis and Time Optimal Control* (Academic Press, New York, London, 1969).

[5] Igor Kluvánek, "The range of a vector-valued measure", *Math. Systems Theory* **7** (1973), 44-54.

[6] Igor Kluvánek and Greg Knowles, *Vector Measures and Control Systems* (North-Holland Mathematics Studies, **20**. Notas de Mátematica, **58**. North-Holland, Amsterdam, Oxford; American Elsevier, New York, 1975).

[7] Gregory Knowles, "Lyapunov vector measures", *SIAM J. Control* **13** (1975), 294-303.

[8] Greg. Knowles, "Some remarks on infinite-dimensional nonlinear control without convexity", *SIAM J. Control and Optimization* **15** (1977), 830-840.

[9] G. Knowles, "Time optimal control of infinite-dimensional systems", *SIAM J. Control and Optimization* **14** (1976), 919-933.

[10] Greg Knowles, "Some problems in the control of distributed systems and their numerical solution", *SIAM J. Control and Optimization* (to appear).

[11] J.P. LaSalle, "Time optimal control systems", *Proc. Nat. Acad. Sci. USA* **45** (1959), 573-577.

[12] David L. Russell, "Boundary value control of the higher-dimensional wave equation", *SIAM J. Control* **9** (1971), 29-42.

[13] David L. Russell, "Boundary value control of the higher-dimensional wave equation, Part II", *SIAM J. Control* **9** (1971), 401-419.

[14] Roberto Triggiani, "Extensions of rank conditions for controllability and observability to Banach spaces and unbounded operators", *SIAM J. Control and Optimization* **14** (1976), 313-338.

[15] A.G. Vituškin, "Coding of signals with finite spectrum and sound recording problems", *Proc. Internat. Congress of Mathematicians*, Vancouver, B.C., 1974, Vol. I, 221-226 (Canadian Mathematical Congress, Montreal, Quebec, 1975).

STATISTICAL FILTERING

John B. Moore

This paper is a tutorial survey which focuses on some developments in statistical filtering achieved since the introduction of Wiener and Kalman filters for linear gaussian problems. Kalman filters (including smoothers and predictors) are reviewed with reference to their interesting properties and also their fundamental limitations in nonlinear or unknown environments. For nonlinear filtering problems, the relevance of the near optimal extended Kalman filters, gaussian sum filters, and bound optimal filters are discussed. For adaptive linear filtering and prediction, connections of the linear gaussian theory with recursive least squares parameter estimation theory are seen to yield adaptive filtering algorithms which are asymptotically optimum, and connections with recursive a posteriori probability updating algorithms are seen to yield optimal solutions to model approximation, fault detection, and adaptive filtering problems.

1. Introduction

Filters were once understood to be simply contrivances for freeing liquids from suspended impurities by passing them through sand or charcoal. Today, with ultra-violet filters on our cameras, and electrical signal filters in our radios we interpret the word *filter* more abstractly. Here we focus on statistical filtering in

---

Work supported by the Australian Research Grants Committee.

which signals or system states are estimated from noisy measurements from knowledge of
the statistics of the signals and unwanted noise signals. Theoretical developments in
statistical filtering have been made side by side with the developments in integrated
circuit, microprocessor, and digital filter technologies. Perhaps in the future more
linkages will be made between these two major developments in filtering.

In the classical approaches to filtering using electrical networks, it is assumed
that the useful signals lie in one frequency band and the unwanted signals (noise) lie
in another with possibly some overlap. In the statistical approach, the best filter
is one which on the average has its output (or states) closest to the correct or
useful signal (or states). The earliest statistical ideas of Wiener [23] and
Kolmogorov [15] in the 1940's relate to processes with statistical properties which do
not change with time; that is, to stationary processes. For these processes it is
possible to relate the statistical properties of the useful signals and unwanted noise
with their frequency domain properties, thus at least conceptually linking classical
and statistical filtering. Both classical and Wiener filters can be implemented using
standard analog circuit components.

In the late 1950's and 1960's, a statistical filtering theory was developed [11,
12, 13, 14] which did not require the stationarity assumption. It arose in order to
cope with certain applications in which the nonstationarity of the signal or noise was
intrinsic to the problem. The resulting filter could be implemented by recursive
algorithms using a digital computer, digital integrated circuit modules, or by time
varying analog networks. The nonstationary statistical theory soon became known as
Kalman filter theory. Since the theory was developed using state space signal models
in the time domain rather than transfer function models in the frequency domain, at
first it seemed that the Wiener and Kalman theories were not closely related, but more
recently a rapprochement is apparent.

The Kalman filter theory assumes known linear and finite dimensional signal
models with input and additive output noise being white noise processes independent of
one another with known means and covariances. The natural criterion of performance
for a filter is the mean of the signal or state estimation error squared conditioned
on the measurements, for when the best linear filter is selected using this criterion,
it turns out to be optimum in the important case when the noise disturbances are
gaussian. The theory and resulting linear filter for the model and criterion noted,
have an attractive simplicity with intuitive appeal, particularly as the online linear
state estimator (filter) equations are decoupled from the matrix Riccati error
covariance equations. The latter must be calculated (off-line) to determine the
crucial filter parameters. The state estimate is a conditional mean estimate
(conditioned on the measurements) and the error covariance is likewise a conditional
error covariance and together, in the gaussian noise case, these yield the conditional
probability density function, or equivalently all the information in the measurements
concerning the desired signals or states.

For the case of nonlinear signal models or linear models with nongaussian noise processes, the conditional probability density functions are no longer gaussian, but one possibility is to somehow extend the linear theory to cope in this case. For the case when the additive noise is low and the initial estimates are good ones, it is likely that the conditional (also termed a posteriori) probabilities will be of low variance and therefore can be approximated by a single gaussian density function. For such cases, it makes sense to linearize the signal model in such a way that a Kalman filter can be constructed for the linear model, keeping in mind that the linear model parameters are now functions of the signal model states and are therefore stochastic. The resulting Kalman filter, termed an extended Kalman filter, has its matrix Riccati error covariance equations coupled to its state estimate equations. It is therefore considerably more complex than the simple Kalman filter to implement, unless some decoupling is forced by the introduction of further assumptions.

A performance theory for a class of such decoupled extended Kalman filters with cone bounded nonlinearities is studied in [6], and in particular performance bounds are given which allow selections of parameters in the filter to optimize these bounds. As the cone bounds tighten, the extended Kalman filter approaches the Kalman filter.

For the case when the a posteriori probability density functions are not of low variance, then they frequently can be adequately represented by the sum of a small number of gaussian terms. It is possible to construct a gaussian sum filter which consists of a bank of extended Kalman filters such that each keeps track of one gaussian term with small variance, and to weight the outputs so as to yield a conditional mean estimate or a maximum a posteriori estimate of the states as illustrated in [21]. The ideas of [6, 21] together yield powerful algorithms for the case when nonlinearities are not tightly cone bounded [7].

A class of nonlinear signal models which is of particular interest are those which can be viewed as linear state models with unknown parameters. To achieve simultaneous parameter and state estimation is of course a nonlinear estimation problem, but to achieve relatively simple estimators, it proves useful to exploit the fact that the unknown parameters are constant or at least slowly varying. One way to do this is to assume that they belong to a discrete set $\{\theta_i\}$ and employ a corresponding set of Kalman filters each conditioned on a different θ_i. An appropriate weighted sum of the state estimates conditioned on the θ_i then yields the desired state estimator. Theories for such estimators are studied in [4, 8]. Another approach is to employ a state estimator conditioned on knowledge of the parameters and a parameter estimator conditioned on knowledge of the states. It then seems reasonable to couple the two estimators so that the true parameters and true states are replaced by the available estimates. A convergence theory which gives conditions for convergence to the optimal estimates is given in [16].

In the next section, the optimal filtering results for the linear gaussian case are reviewed. In Section 3, the asymptotically optimal adaptive filtering results are presented and in Section 4 the near optimal filtering results for nonlinear signal models are discussed.

2. Linear optimal filtering

In order to have a filtering problem there must be a signal generating *system* which is usually physical and thus causal and dynamic. The system may operate in discrete or continuous time with underlying difference or differential equations and the output may change at discrete-time instants or on a continuous basis. It is also implicit that the systems are noisy in that the inputs may be unpredictable except for their statistical properties and the outputs may be derived from a sensor with random inaccuracies or after passage over a noisy channel.

The term *filtering* implies an estimation of some quantity of the system which we would like to know based on the measurements. We may like to know the states $\{x_k\}$ in a system we need to control or the signal $\{y_k\}$ in a communication system. More specifically filtering is used to denote estimation of x_k (say) given measurements $z_0, z_1, \ldots, z_k$, *smoothing* denotes estimation of x_k given measurements $z_0, z_1, \ldots, z_{k+N}$ for some $N > 0$, and *prediction* denotes estimation of x_k given $z_0, z_1, \ldots, z_{k+N}$ for some $N < 0$. Filtering is used in radio reception, smoothing is used to read bad writing, and prediction is used to catch a ball.

A crucial notion in statistical filtering theory is to define an estimation criterion and select a filter which by this criterion is optimum. It is usual to base the criterion on some feature of the a posteriori probability density $p(x_k \mid z_0, z_1, \ldots, z_{k+N})$ for some N. This density tells us all we know of x_k given the measurements $z_0, z_1, \ldots, z_{k+N}$. The *maximum a posteriori* estimate maximizes this density, and the *conditional mean estimate* $\hat{x}_{k|k+N} = E[x_k \mid z_0, z_1, \ldots, z_{k+N}]$ turns out to be the unique conditional *minimum variance estimate* $\hat{x}_k$ such that

$$E\{\|x_k - \hat{x}_k\|^2 \mid z_0, z_1, \ldots, z_{k+N}\} \leq E\{\|x_k - x_k^*\|^2 \mid z_0, z_1, \ldots, z_{k+N}\}$$

for all vectors x_k^*. This latter estimate has strong intuitive appeal, but is clearly the same as the maximum a posteriori estimate when the a posteriori distribution is symmetrical and unimodal as in the case of linear systems and gaussian noise when the a posteriori is in fact gaussian.

The power of the optimal filtering approach is that for linear signal models, the best linear filter which is known as the Kalman filter has very useful properties. Let us consider the *linear signal model* with state space equations

$$x_{k+1} = F_k x_k + w_k,$$

$$z_k = y_k + v_k = H'_k x_k + v_k,$$

$$E\begin{bmatrix} w_k \\ v_k \end{bmatrix} = 0, \quad E\left\{ \begin{bmatrix} w_k \\ v_k \end{bmatrix} [w'_l \; v'_l] \right\} = \begin{bmatrix} Q_k & 0 \\ 0 & R_k \end{bmatrix} \delta_{kl},$$

$$E[x_0] = \bar{x}_0, \quad E[(x_0 - \bar{x}_0)(x_0 - \bar{x}_0)'] = P_0,$$

where the subscript is a time argument, $\{x_k\}$ is the system state vector, $\{v_k\}$ is the white measurement noise process, $\{w_k\}$ is the white input vector (assumed independent of v_k) and $\{z_k\}$ is the output vector process. The matrices F_k, H_k, Q_k, R_k, P_0 are assumed known *a priori*.

The *Kalman filter* [11, 12, 13, 14] gives the one step ahead prediction estimate $\hat{x}_{k/k-1}$ for the above signal model in the gaussian noise case, and is the best linear filter, in that the unconditioned error covariance is minimized, otherwise. Denoting the state estimate as $\hat{x}_k$, which is the one step ahead prediction estimate $\hat{x}_{k/k-1}$ in the gaussian case, the filter equations are

$$\hat{x}_{k+1} = F_k \hat{x}_k + K_k (z_k - H'_k \hat{x}_k), \quad \hat{x}_0 = \bar{x}_0,$$

where the Kalman gain K_k is calculated in terms of the error covariance matrix $\Sigma_{k/k-1} = E[(x_k - \hat{x}_k)(x_k - \hat{x}_k)' \mid z_0, z_1, \ldots, z_k]$ as

$$K_k = F_k \Sigma_{k/k-1} H_k [H'_k \Sigma_{k/k-1} H_k + R_k]^{-1},$$

$$\Sigma_{k+1/k} = F_k \left[\Sigma_{k/k-1} - \Sigma_{k/k-1} H_k [H'_k \Sigma_{k/k-1} H_k + R_k]^{-1} H'_k \Sigma_{k/k-1} \right] F'_k + Q_k.$$

There are a number of *properties* of the above linear filter which are interesting from a theoretical point of view and an understanding of them is important for many applications. These properties and general comments about the Kalman filter and derivations of the filter equations are now listed.

1. A first principles derivation of the discrete-time Kalman filter when the noise disturbances are gaussian is readily achieved using a step by step application of the well known result that if X and Y are jointly gaussian with $Z = [X' \; Y']$ possessing mean and covariance

$$\begin{bmatrix} \bar{x} \\ \bar{y} \end{bmatrix} \quad \text{and} \quad \begin{bmatrix} \Sigma_{xx} & \Sigma_{xy} \\ \Sigma_{yx} & \Sigma_{yy} \end{bmatrix},$$

then X conditioned on the information that $Y = y$ is gaussian with mean $\bar{x} + \Sigma_{xx}\Sigma_{xy}^{-1}(y-\bar{y})$ and covariance $\Sigma_{xx} - \Sigma_{xy}\Sigma_{yy}^{-1}\Sigma_{yx}$. The proof is quick and easy and allows F_k, H_k to be stochastic as when they are functions of $z_0, z_1, \ldots, z_{k-1}$. However the proof does not permit a relaxation of the gaussian noise assumption or of the assumption that v_k and w_k are independent, nor does it yield insight into the whitening filter property discussed below.

2. The filter is fortuitously linear, discrete-time, finite-dimensional and readily implemented on a digital computer.

3. The state estimates x_k depend of course on $z_0, z_1, \ldots, z_{k-1}$, but in the usual case when F_k, H_k are deterministic, the conditional error covariance $\Sigma_{k/k-1}$ is in fact independent of $z_0, z_1, \ldots, z_{k-1}$ as is the Kalman gain K_k. The calculation of $\Sigma_{k/k-1}$ and K_k can be carried out off-line. When F_k, H_k are functions of the measurements, then the equations are coupled and calculation of $\Sigma_{k/k-1}$ and K_k must be carried out on-line at considerable expense.

4. The filter may be viewed as consisting of a model of the system with a feed-back controller K_k which feeds back the error $(z_k - H_k'\hat{x}_k)$ being the difference between the actual system output and the one-step ahead predicted output. The lower is the gain K_k the less is the effect that the measurements have on the estimates and so the greater is the noise filtering, and the higher the gain K_k the less the error $(z_k - H_k'\hat{x}_k)$ when noise is absent in the measurement generating process.

5. When the noise process $\{v_k\}$ and $\{w_k\}$ are gaussian, then $\{x_k\}$ and $\{z_k\}$ are jointly gaussian and $\hat{x}_k = \hat{x}_{k/k-1}$ is also gaussian. Thus $\hat{x}_{k/k-1}$ and $\Sigma_{k/k-1}$ together yield $p(x_k \mid z_0, \ldots, z_{k-1})$.

6. Should $(H_k'\Sigma_{k/k-1}H_k + R_k)$ become singular, its inverse may be replaced in the calculations by its pseudo inverse.

7. If F_k, H_k, Q_k, R_k are in fact time-invariant and the underlying processes are stationary, the filter gain K_k will still be time-varying. Under assumptions of complete detectability and stabilizability, the filter is asymptotically time invariant and stable for arbitrary initial error covariance. In the case when the initial time is $-\infty$, requiring the signal model to be asymptotically stable, the

filter is time-invariant and asymptotically stable. Moreover, with z the z-transform variable and $\Sigma_{k/k-1} = \Sigma$ the following frequency domain formula is valid:

$$[I+H'(zI-F)K][R+H'\overline{\Sigma}H]\left[I+K'\left(z^{-1}I-F\right)^{-1}H\right] = R + H'(zI-R)^{-1}Q\left(z^{-1}I-F'\right)^{-1}H ,$$

where $H'(zI-F)^{-1} = W_m(z)$ is the transfer function of the signal process model and $R + W_m(z)QW_m'(z^{-1})$ is the power spectrum $\Phi(z)$ of the output process $\{z_k\}$. Defining $W_K(z)$ to be $H'(zI-F)^{-1}K$ the above equation can be rewritten as

$$[I+W_K(z)][R+H'\overline{\Sigma}H]\left[I+W_K'\left(z^{-1}\right)\right] = R + W_m(z)QW_m'\left(z^{-1}\right) ,$$

which is in essence a spectral factorization of $\Phi(z)$ as $\Phi(z) = W(z)W'\left(z^{-1}\right)$ where $W(z) = \left[I+H'(zI-F)^{-1}K\right](H'\overline{\Sigma}H+R)^{\frac{1}{2}}$ is a minimum phase spectral factor. The filter transfer function with output $\hat{y}_k = H'\hat{x}_k$ is

$$W_f(z) = H'(zI-\overline{F-KH'})^{-1}K$$

$$= W_K(z)\left[I+W_K(z)\right]^{-1} .$$

Thus the filter transfer function is related to the signal power spectrum and can be derived from it by classical procedures.

8. The *innovations* sequence $\tilde{z}_k = z_k - H_k'\hat{x}_k$ consists of that part of z_k containing new information not carried in $z_{k-1}, z_{k-2}, \ldots$ in that $\tilde{z}_k = z_k - E[z_k \mid z_0, z_1, \ldots, z_{k-1}]$ in the event that $\tilde{z}_k$ is gaussian. It is not difficult to show that $\tilde{z}_k$ is a linear function of $z_0, z_1, \ldots, z_k$ and is orthogonal to $z_0, z_1, \ldots, z_{k-1}$ by the orthogonality principle and therefore that $\tilde{z}_k$ is *white* in that $E[\tilde{z}_k \tilde{z}_l'] = 0$ for $k \neq l$. The Kalman filter driven by $\{z_k\}$ and with output $\tilde{z}_k$ is thus a whitening filter. It is also of interest that $E[x_k \mid z_0, z_1, \ldots, z_{k-1}] = E[x_k \mid \tilde{z}_0, \tilde{z}_1, \ldots, \tilde{z}_{k-1}]$ holds. For the nongaussian case, the above ideas go through with $E[z_k \mid z_0, z_1, \ldots, z_{k-1}]$ replaced by a linear minimum variance estimate denoted $E^*[z_k \mid z_0, z_1, \ldots, z_{k-1}]$ which as we have noted is the conditional mean estimate $E[z_k \mid z_0, z_1, \ldots, z_{k-1}]$ in the gaussian case. It is also an interesting property that the linear minimum variance estimate is unbiased in that $E\{E^*[z_k \mid z_0, z_1, \ldots, z_{k-1}]\} = E[z_k]$. Perhaps the most instructive of the derivations of the Kalman filter first develops these innovations ideas and then applies the projection theorem as in [3].

9. The addition of a known external input u_k to the signal model simply requires the addition of the same external input to the filter.

10. The filter input signal to noise ratio can be defined as $[H'PH]R^{-1}$ and the filter output signal to noise ratio as $H'(P-\Sigma)H[H'\Sigma H]^{-1}$, at least for the time-invariant filter case, and it is readily shown that there is always a signal to noise ratio improvement introduced by the Kalman filter [3].

11. Suppose there is given a filter with the feedback structure of the Kalman filter with the term corresponding to the innovations as $\{z_k - q_k\}$ for some q_k which is $z_0, z_1, \ldots, z_{k-1}$ measurable and where z_k is recoverable from $\{z_l - q_l, l \le k\}$, then $q_k = \hat{z}_{k/k-1}$ if and only if $\{z_k - q_k\}$ is zero mean and white. The whiteness is implied by the condition $E\{[z_{k+l} - q_{k+l}][z_k - q_k]'\} = 0$ for all $0 < l \le m+n$ where m and n are the dimensions of the signal model and filter states respectively. We thus have a test for optimality of the filter as a signal filter. The addition of certain assumptions on the co-ordinate basis and observability lead to the same conditions for optimality of the filter as a state estimator [3].

12. If the covariances Q_k, R_k, P_0 are unknown but the design values Q_k^d, R_k^d, P_0^d are taken as being conservative estimates in that $Q_k^d \ge Q_k$, $R_k^d \ge R_k$, $P_0^d \ge P_0$, then the design error covariance matrix $\Sigma_{k/k-1}^d$ is a conservative estimate of the actual error covariance, denoted $\Sigma_{k/k-1}^a$, achieved using the designed filter in that $\Sigma_{k/k-1}^d \ge \Sigma_{k/k-1}^a \ge \Sigma_{k/k-1}$, [9]. Again if errors in any parameter are possible, then to guarantee a bound on $\Sigma_{k/k-1}^a$ one almost always needs exponential asymptotic stability of the actual signal generating system [10]. When $\Sigma_{k/k-1}^d$ remains bounded but $\Sigma_{k/k-1}^a$ diverges, then the filter can usually be tamed by increasing Q_k^d, R_k^d and P_0^d and thereby K_k^d. A convenient method of achieving this is known as exponential data weighting [1] which also achieves a prescribed degree of stability α in the filter simply by replacing F_k by $F_k^d = \alpha F_k$.

13. Minor variations to the recursive relationships yield the true filtered estimates $\hat{x}_{k/k}$ and $\Sigma_{k/k}$ and also allow $E[w_k v_l'] = S\delta_{kl}$ for some nonzero S. Alternative formulations, *the information filter*, update $\Sigma_{k/k}^{-1}$ or $\Sigma_{k/k-1}^{-1}$ and $\hat{a}_{k/k-1} = \Sigma_{k/k-1}^{-1} \hat{x}_{k/k-1}$ or $\hat{a}_{k/k} = \Sigma_{k/k}^{-1} \hat{x}_{k/k}$ and for some problems the calculations are

more efficient. Updating the square root of $\Sigma_{k/k}$ or $\Sigma_{k/k}^{-1}$ leads to what are termed *square root filtering* algorithms which are frequently more reliable when Σ is small [5]. Also with R_k a diagonal or block diagonal matrix there is some computational advantages in processing the vector outputs sequentially [20]. In the *high measurement noise* case, simplification is possible by simply setting $[H\Sigma H+R]^{-1} = 0$ in the calculation of Σ [3]. For the time invariant filter case there are computational efficiencies which can be achieved using Chandresekhar [18], and doubling algorithms [2].

14. *Prediction* is readily achieved using a Kalman filter since via the signal model state equation x_{k+N} can be expressed in terms of x_k and other terms to yield $\hat{x}_{k+N/k} = \prod_{i=1}^{N} F_{k+N-i} \hat{x}_{k/k}$. Perhaps the most useful form of smoothing is *fixed-lag smoothing*. The estimate $\hat{x}_{k-N/k}$ improves as N increases with a time constant roughly given by the dominant time constant of the Kalman filter and thus for N chosen as two or three times this value there is achieved virtually all the smoothing that can be achieved. The fixed-lag smoother is simply a Kalman filter for the signal model above augmented with delays in the states so that $d_k = x_{k-N}$ is a state of the signal model, for then $\hat{d}_{k/k} = \hat{x}_{k-N/k}$ [17].

We see that indeed there is a rich theory for optimal linear filtering, but let us be aware that most real world signal generating systems are non-finite-dimensional, nonlinear, and even, if they can be adequately modelled by a linear system, there is uncertainty as to what the parameters should be. Moreover, the criterion of performance that we have been using could well be inappropriate.

3. Adaptive filtering

When the scientific laws of the signal generating system are obscure or too complex to yield precise signal model descriptions, time-series analysis techniques can be employed to yield either output covariances and from these a signal model, or a filter directly. These techniques are not explored here, rather adaptive filtering algorithms are discussed. Their appeal lies in an ability to accommodate to slowly varying signal generating system parameter changes in an on-line calculation.

In this section we restrict attention to linear signal models with unknown parameters which may be slowly time-varying or time invariant and consider the simultaneous estimation of model states and parameters. It is usual not to view the unknown parameters as additional states in a highly nonlinear estimation problem since optimal estimation in this case is too formidable an undertaking, but rather the fact that the parameters are constant or nearly so is exploited to achieve asymptotically optimum or near optimum adaptive identification and state estimation schemes. For simplicity, the parameters are assumed constant and one parallel processing scheme and one extended least squares algorithm are discussed.

Parallel Processing Scheme. Let us assume that the unknown parameter θ belongs to a discrete set $\{\theta_1, \theta_2, \ldots, \theta_N\}$ with assumed *a priori* probabilities for each θ_i. Then the conditional mean estimate is

$$\hat{x}_{k/k-1} = \sum_{i=1}^{N} \hat{x}_{k/k-1, \theta_i} p_{\theta_i/k-1},$$

where $\hat{x}_{k/k-1, \theta_i}$ are state estimates conditioned on θ_i and $p_{\theta_i/k-1}$ denotes the a posteriori probability for θ_i. The state estimates can be calculated recursively via a bank of Kalman filters and the conditioned probabilities can be calculated recursively from [19],

$$p_{\theta_i/k} = C \left| \Omega_{k/\theta_i}^{-1} \right|^{\frac{1}{2}} \exp\left\{ -\tfrac{1}{2} \tilde{z}'_{k/\theta_i} \Omega_{k/\theta_i}^{-1} \tilde{z}_{k/\theta_i} \right\} p_{\theta_i/k-1},$$

where C is a normalizing constant independent of θ_i, the residual $\tilde{z}_{k/\theta_i} = z_k - \hat{z}_{k/k-1, \theta_i}$ is given from the conditioned Kalman filters, and $\Omega_{k/\theta_i} = E[\tilde{z}_{k/\theta_i} \tilde{z}'_{k/\theta_i}]$ is given in advance from the conditioned filter covariance calculations.

Convergence results are available [4] which give very reasonable conditions for (exponential) convergence of $p_{\theta_i/k}$ to 1 for the case when θ_T is the true θ. A convenient but not a necessary condition is asymptotic ergodicity of $\tilde{z}_{k/\theta_i}$ in its autocorrelation. There is also needed an identifiability condition that for $\theta_i \neq \theta_j$, either $\tilde{z}_{k/\theta_i} - \tilde{z}_{k/\theta_j}$ fails to approach zero, or $\Omega_{/k} \neq \Omega_{j/k}$ as $k \to \infty$ or both. If the z_k are gaussian then the convergence is with probability one.

The results as stated above are immediately useful in fault detection where the θ_i correspond to fault conditions [24].

For the case when the unknown parameters belong to an infinite set, it makes sense to use the same adaptive estimator as would be used when the set is finite. Convergence results in [4] tell us that there will still be exponential convergence of $p_{\theta_j/k}$ to 1 for one of the θ_i, denoted θ_j, under similar conditions to those noted above, and moreover θ_j is nearest the true θ_T (assumed not to belong to the finite set $\theta_1, \theta_2, \ldots, \theta_N$) in the following sense. Defining $\beta_i = \ln|\Omega_i| + \text{tr}\left(\Omega_i^{-1}\Sigma_i\right)$ where it is assumed that $\Omega_{k/\theta_i} \to \Omega_i$ as $k \to \infty$ and $\lim_{k\to\infty} n^{-1} \sum_{j=k}^{k+n-1} \tilde{z}_{j/\theta_i} \tilde{z}'_{j/\theta_i} = \Sigma_i$, then $\beta_j < \beta_i$ for all $i \neq j$, and the exponent of the exponential convergence of $\dfrac{p_{\theta_i/k}}{p_{\theta_j/k}}$ to zero is $(\beta_j - \beta_i)/2$. There is a close relationship of the measure β_i to the Kullback information measure and also to spectrum measures [4]. In selecting the finite set of θ_i, it becomes clear that they should be evenly spaced in the sense of the β_i measures.

The results for the case when θ is not in a finite set are immediately useful for the case of selecting a low order model most suited to represent a high order system.

Extended Least Squares Adaptive Filtering. The parallel processing scheme above may be too costly to implement and simpler schemes may perform almost as well in certain applications. A very useful class of adaptive estimators arises if it is possible to calculate both the conditional mean state estimate $\hat{x}_{k/k-1}$, and the least squares parameter estimates $\theta^*_{k/k-1, x_{k-1}}$.

In the spirit of extended least squares algorithms, we can replace θ and x_{k-1} by the estimates of these, in the conditioning variables of the above estimates which will clearly lead to suboptimal state and parameter estimates, denoted $\hat{x}_k$ and $\hat{\theta}_k$. Such estimates are very useful in some situations since they are asymptotically optimum.

Signal models for which the extended least squares notions above can be applied require measurements of the form $z_k = y_k + v_k$ where $y_k = \theta' x_k$ and v_k is a process of independent variables. The states x_k are assumed to arise from a model $x_{k+1} = \bar{F} x_k + \bar{G} y_k + \bar{K} v_k + \bar{B} z_k + u_k$ for some known $\bar{F}, \bar{G}, \bar{K}, \bar{B}, u_k$ (possibly time varying). The resulting estimator equations are

$$\hat{x}_{k+1} = \bar{F}\hat{x}_k + \bar{G}\hat{y}_k + \bar{K}\hat{v}_k + \bar{B}z_k + u_k,$$

$$\hat{v}_k = z_k - \hat{y}_k, \quad \hat{y}_k = \hat{\theta}'_k \hat{x}_k,$$

$$\hat{\theta}_k = \hat{\theta}_{k-1} + \Lambda_{k+1}\hat{x}_k(z'_k - \hat{x}'_k \hat{\theta}_{k-1}),$$

$$\Lambda_{k+1} = \Lambda_k - \Lambda_k \hat{x}_k(\hat{x}'_k \Lambda_k \hat{x}_k + 1)^{-1} \hat{x}'_k \Lambda_k.$$

Convergence theory [16], tells us that with

$$W(z) = \tfrac{1}{2}I + \theta'\{zI-[\overline{F}+(\overline{G}-\overline{K})\theta']\}(\overline{G}-\overline{K})$$

strictly positive real and with reasonable persistently exciting conditions on x_k, very like those required in standard least squares, satisfied then $\hat{x}_k \to \hat{x}_{k/k-1,\theta}$ and $\hat{\theta}_k \to \theta^*_{k/k-1,x_{k-1}}$ almost surely as $k \to \infty$.

This theory is merely a guideline which gives some degree of confidence in the approach. Many simulation examples and some applications have confirmed the usefulness of the algorithms.

The above theory specializes to yield results for autoregressive moving average parameter identification, model reference adaptive schemes and other schemes but here we note in particular an application to the case of adaptively adjusting the Kalman gain K of a Kalman filter applied in a situation where the noise variances Q, S, R in the signal model of Section 2 are unknown, but the parameters reflecting the model dynamics F and H are known. To keep the ideas simple, let us consider the case of scalar z_k and note that $H'(zI-F)^{-1}K = K'(zI-F')^{-1}H$. The appropriate adaptive filter can then be seen to have equations as above but with $\overline{F}, \overline{G}, \overline{K}, \overline{B}, u_k$ and $\overline{\theta}$ replaced by $F', 0, H, 0, 0$, and K respectively. The positive real condition is simply that $\tfrac{1}{2}I - K'[zI-(F'-HK')]^{-1}H$ be positive real. This condition is satisfied should the Kalman filter be asymptotically stable and for the noise covariance R of the signal model sufficiently large, at least in the usual case when S of the signal model is zero.

Recursive maximum likelihood schemes can also be devised, being mildly more sophisticated than those described so far and in some instances it appears with less restrictions in the convergence conditions.

4. Nonlinear filtering

We have seen that in the special class of nonlinear filtering problem known as adaptive filtering where some of the states (termed parameters) are constant, that the ideal of optimal filtering has to be relaxed and we are fortunate if near optimal filtering or asymptotically optimal filtering can be achieved. For more general nonlinear filtering, again we hover near optimality if possible, perhaps by restricting the noise, or the nonlinearities, or by tolerating filters which demand large amounts of computational effort. To limit the discussion in this section, three nonlinear filtering schemes which arise in some way from the linear theory will be considered.

Extended Kalman Filters. So as not to depart too far from the linear gaussian model consider the system

$$x_{k+1} = f_k(x_k) + g_k(x_k)w_k,$$

$$z_k = h_k(x_k) + v_k,$$

with

$$F_k = \left.\frac{\partial f_k(x)}{\partial x}\right|_{x=\hat{x}_{k/k}}, \quad H'_k = \left.\frac{\partial h_k(x)}{\partial x}\right|_{x=\hat{x}_{k/k-1}} \quad \text{and} \quad G_k = g_k(\hat{x}_{k/k}).$$

That is, neglecting higher order terms in a Taylor series expansion of $f_k(x_k)$, and so on, we have an approximate model

$$x_{k+1} = F_k x_k + G_k w_k + u_k,$$

$$z_k = H'_k x_k + v_k + y_k,$$

where u_k and y_k are calculated on line from the equations

$$u_k = f_k(\hat{x}_{k/k}) - F_k \hat{x}_{k/k}, \quad y_k = h_k(\hat{x}_{k/k-1}) - H'_k \hat{x}_{k/k-1}.$$

The Kalman filter for this approximate (linearized) model is a trivial variation of that given in Section 2 and is known as the extended Kalman filter for the nonlinear model. Its equations are

$$\hat{x}_{k/k} = \hat{x}_{k/k-1} + L_k[z_k - h_k(\hat{x}_{k/k-1})],$$

$$\hat{x}_{k+1/k} = f_k(\hat{x}_{k/k}),$$

$$L_k = \Sigma_{k/k-1} H_k (H'_k \Sigma_{k/k-1} H_k + R_k)^{-1},$$

$$\Sigma_{k/k} = \Sigma_{k/k-1} - \Sigma_{k/k-1} H'_k (H'_k \Sigma_{k/k-1} H_k + R_k)^{-1} H_k \Sigma_{k/k-1},$$

$$\Sigma_{k/k} = F_k \Sigma_{k/k} F'_k + G_k Q_k G'_k,$$

with initializations $\Sigma_{0/-1} = P_0$, $\hat{x}_{0/-1} = \bar{x}_0$.

This Kalman filter is clearly optimal for the approximate model but is inevitably suboptimal when applied to the original nonlinear model. As a consequence, the notations $\hat{x}_{k/k}$ and $\Sigma_{k/k-1}$ are now loose denoting approximate estimates and covariances. The smaller is $\|x_k - \hat{x}_{k/k}\|^2$ and $\|x_k - x_{k/k-1}\|^2$, as in high signal to noise ratio situations with good initial estimates, the better are the approximations. The on-line calculations for $\operatorname{tr} \Sigma_{k/k}$ and $\operatorname{tr} \Sigma_{k/k-1}$ or the whiteness of the pseudo-innovations could give a good idea as to how nearly optimal the filter is.

One theoretical insight [3], tells us that if $p(x_k \mid z_0 \ldots z_{k-1})$ is gaussian with mean $x_k - \hat{x}_{k/k-1}$ and covariance $\Sigma_{k/k-1}$, then as $\Sigma_{k/k} \to 0$,

$p(x_k \mid z_0 \ldots z_k)$ converges uniformly to a gaussian density with mean $\hat{x}_{k/k}$, and covariance $\Sigma_{k/k}$. Likewise with the conditioning variables $z_0, \ldots, z_k$ with appropriate notational changes.

The covariance and estimate equations are clearly coupled and must both be calculated on line. Forcing the Kalman gain to be independent of the states is desirable from a computational viewpoint. An illustration of this is to the near optimal demodulation of FM signals in high noise. Scalar sampling leads to coupled filter equations and when decoupling is forced there results the familiar phase locked loop. With quadrature and in phase sampling, however, the filter equations are fortuitously not coupled and the performance of this filter is as good as the coupled filter for the scalar sampling case and represents a significant improvement on the standard phase locked loop [22].

In practice, the selection of a signal model co-ordinate basis may be crucial to achieving an extended Kalman filter with good stability properties. A study using the Lyapunov function $\hat{x}'_{k/k} \Sigma_{k/k}^{-1} \hat{x}_{k/k}$ for the autonomous system can be used to select an appropriate co-ordinate basis. Sometimes it proves helpful to add dither to the filter nonlinearities to improve the stability properties. Case studies are yielding more and more useful tricks to improve the performance of the extended Kalman filter. The obvious idea to use higher order approximations in the Taylor series expansions tends to yield more complex filters with dubious performance improvement, at least beyond the second order versions.

Bound Optimum Filters. For the case when $f_k(x_k)$ and $h_k(x_k)$ in the above nonlinear signal model are cone-bounded and $g_k(x_k)$ is assumed to be G_k, then forcing decoupling of the extended Kalman filter equations simply by setting the Kalman gain L_k to be selected *a priori* allows calculations of performance bounds from the cone bounds.

Consider the cone bounds

$$\|f_k(x+\delta x) - f_k(x) - \overline{F}_k \delta x\| \leq \|\Delta \overline{F}_k x\| \delta ,$$

$$\|h_k(x+\delta x) - h_k(x) - \overline{H}'_k \delta x\| \leq \|\Delta \overline{H}'_k x\| \delta ,$$

for all x and scalar δ, and some matrices $\overline{F}_k, \Delta \overline{F}_k$, and so on, independent of x and δ. Somewhat tedious manipulations [3] now show that these cone bounds imply the performance bounds for the filter above as

$$\Sigma_{k/k} \leq \overline{\Sigma}_{k/k} , \quad \Sigma_{k+1/k} \leq \overline{\Sigma}_{k+1/k}$$

where for arbitrary α and β and initializations $\overline{\Sigma}_{0/-1} = \Sigma_{0/-1}$,

$$\overline{\Sigma}_{k+1/k} = (1+\alpha)\overline{F}_k\overline{\Sigma}_{k/k}\overline{F}_k' + (1+\alpha^{-1})\mathrm{tr}\left(\Delta\overline{F}_k\overline{\Sigma}_{k/k}\Delta\overline{F}_k'\right)I + G_kQ_kG_k \;,$$

$$\overline{\Sigma}_{k/k} = (1+\beta)\left(I-L_k\overline{H}_k'\right)\overline{\Sigma}_{k/k-1}\left(I-L_k\overline{H}_k'\right)' + L_kR_kL_k' + (1+\rho^{-1})\mathrm{tr}\left(\Delta\overline{H}_k'\overline{\Sigma}_{k/k-1}\Delta\overline{H}_k\right)L_kL_k' \;.$$

These equations for $\overline{\Sigma}$ are virtually standard error covariance Riccati equations for some linear model case and as $\Delta\overline{H}$ and $\Delta\overline{F}$ approach zero (the cone bounds force linearity) then the standard linear filtering equations are recovered. Thus we have imbedded the linear filter in a class of nonlinear ones and thereby gained insight into what happens when the model is approximately linear in a cone bound sense.

Clearly L_k can be selected so as to minimize the bounds $\overline{\Sigma}$, but the details [3, 6] are omitted here. The important point to note is that when nonlinearity is introduced into the model then filters which are relatively simple to build and correspond to ready extensions of the linear filters cannot be made optimum in the minimum square error sense but can be made to optimize a bound on this index. The more nearly linear the model, the tighter the bound.

For the case when the nonlinearities are far from linear, it may be possible to set up a combined detection and estimation problem in which at each time instant the nonlinearities can be modelled by tight cone bounds. The estimation scheme in this case consists of a bank of bound optimum cone-bounded filters with outputs weighted according to the conditional probabilities for each. That is, the ideas of adaptive estimation via parallel processing of the previous section and the bound optimal filter ideas are generalized to yield useful results for a highly nonlinear signal model case [7].

Gaussian Sum Estimators. When the complexity of a nonlinear filter is permitted to increase beyond that of an extended Kalman filter, it makes sense to try to calculate the a posteriori probability density functions or sufficient statistics of these. An approach to do this which reduces to the extended Kalman filter as the density covariance becomes small requires the representation of the density as the sum of gaussian terms each with a small covariance - the more terms the better the approximation.

A gaussian sum estimator consists of:

1. A bank of extended Kalman filters each of which yields a near optimal estimate of the mean and covariance of one of the gaussian terms in the gaussian sum representation of the a posteriori density.

2. A recursive algorithm which updates the weights of each of the gaussian terms from calculations on the pseudo innovations.

3. An algorithm for converting the means, covariances and weights given from (1), (2) for each of the gaussian terms of the gaussian sum representation of the conditional density into another such set giving as close as possible a density to the

conditional density but with reduced covariances in the gaussian terms. This algorithm is applied either periodically or whenever the evolution of one of the covariances associated with an extended Kalman filter is such that if it grows any larger then it is no longer operating near optimally.

In the limit as the number of gaussian terms (the numbers of extended Kalman filters in the filter bank) approaches infinity, then it can be shown that the gaussian sum filter becomes optimal. For the example of [21] in which FM signals in high noise are filtered, simplifications give a scheme which yields virtually optimum performance with only six filters in the filter bank. The addition of more gives negligible improvement. Of course even so the number of situations in which the effort could be justified by the 5 or 6 db signal to noise ratio threshold extensions in the FM example would be small.

Clearly, the gaussian sum filter is considerably more sophisticated than an extended Kalman filter, but as one case study [21] indicates, it can be appreciably more efficient than more brute force attempts to solve the partial differential equations which describe the evolution of the a posteriori density.

References

[1] Brian D.O. Anderson, "Exponential data weighting in the Kalman-Bucy filter", *Information Sci.* 5 (1973), 217-230.

[2] B.D.O. Anderson, "Second-order convergent algorithms for the steady-state Riccati equation", *Internat. J. Control* (to appear).

[3] B.D.O. Anderson and J.B. Moore, *Optimal Filtering* (Prentice Hall, to appear 1978).

[4] B.D.O. Anderson, J.B. Moore and R.M. Hawkes, "Model approximations via prediction error identification", submitted.

[5] G.J. Bierman, "A comparison of discrete linear filtering algorithms", *IEEE Trans. Aerospace and Electron. Systems* AES-9 (1973), 28-37.

[6] Alfred S. Gilman and Ian B. Rhodes, "Cone-bounded nonlinearities and mean-square bounds - estimation upper bound", *IEEE Trans. Automatic Control* AC-18 (1973), 260-265.

[7] R.M. Hawkes and J.B. Moore, "Analysis of detection - estimation algorithm using cone-bounds", *Proceedings of International Conference on Information Sciences*, (Patras, Greece, 1975).

[8] Richard M. Hawkes and John B. Moore, "Performance bounds for adaptive estimation", *Proc. IEEE* 64 (1976), 1143-1150.

[9] H. Heffes, "The effect of erroneous models on the Kalman filter response", *IEEE Trans. Automatic Control* AC-11 (1966), 541-543.

[10] A.H. Jazwinski, *Stochastic Processes and Filtering Theory* (Academic Press, New York and London, 1970).

[11] Thomas Kailath, "A view of three decades of linear filtering theory", *IEEE Trans. Information Theory* IT-20 (1974), 146-181.

[12] R.E. Kalman, "A new approach to linear filtering and prediction problems", *Trans. ASME Ser. D J. Basic Engrg.* 82 (1960), 35-45.

[13] R.E. Kalman, "New methods in Wiener filtering theory", *Proc. Symp. Eng. Appl. Random Functions Theory and Probability*, (John Wiley & Sons, New York, 1963).

[14] R.E. Kalman and R.S. Bucy, "New results in linear filtering and prediction theory", *Trans. ASME Ser. D J. Basic Engrg.* 83 (1961), 95-108.

[15] A.N. Kolmogorov, "Interpolation und extrapolation von stationären zufälligen Folgen", *Bull. Acad. Sci. URSS Sér. Math. [Izv. Akad. Nauk SSSR]* 5 (1941), 3-14.

[16] G. Ledwich and J.B. Moore, "Multivariable adaptive parameter and state estimators with convergence analysis", submitted.

[17] John B. Moore, "Discrete-time fixed-lag smoothing algorithms", *Automatica - J. IFAC* 9 (1973), 163-173.

[18] Martin Morf, Gursharan S. Sidhu and Thomas Kailath, "Some new algorithms for recursive estimation in constant, linear, discrete-time systems", *IEEE Trans. Automatic Control* AC-19 (1974), 315-323.

[19] F.L. Sims, D.G. Lainiotis and D.T. Magill, "Recursive algorithm for the calculation of the adaptive Kalman filter weighting coefficients", *IEEE Trans. Automatic Control* AC-14 (1969), 215-218.

[20] Robert A. Singer and Ronald G. Sea, "Increasing the computational efficiency of discrete Kalman filters", *IEEE Trans. Automatic Control* AC-16 (1971), 254-257.

[21] P.K. Tam and J.B. Moore, "A gaussian sum approach to phase and frequency estimation", *IEEE Trans. Comm.* (to appear).

[22] P.K. Tam and J.B. Moore, "Improved demodulation of sampled - FM signals in high noise", *IEEE Trans. Comm.* (to appear).

[23] Norbert Wiener, *Extrapolation, Interpolation, and Smoothing of Stationary Time Series* (The Technology Press of M.I.T.; John Wiley and Chapman Hall, London; 1949).

[24] A. Willsky, "A generalized likelihood ratio approach to state estimation in linear systems subject to abrupt changes", *Proc. IEEE* 1974 *Dec. and Contr. Conference*, 846-853 (Phoenix, Arizona, 1974).

SINGULAR PERTURBATIONS AND OPTIMAL CONTROL

R.E. O'Malley, Jr.

These lecture notes are intended to provide an elementary account of some of the recent mathematical effort in applying singular perturbations theory to optimal control problems, to demonstrate the practical importance of this asymptotic technique to current engineering studies, and to suggest several open problems needing further research. Readers are referred to the survey article by Kokotovic, O'Malley, and Sannuti for a discussion of related topics and for additional references.

1. A Simple Singular Perturbation Problem

Following Cole [18], we consider the motion of a linear oscillator initially at rest, subject to an impulse of strength I_0. To find the displacement y, we need to solve the initial value problem

(1) $$m\frac{d^2y}{dt^2} + \beta \frac{dy}{dt} + ky = I_0 \delta(t), \quad y(0^-) = \frac{dy}{dt}(0^-) = 0,$$

where m, β, and k are the usual mass, damping, and spring constants and $\delta(t)$ is a delta function peaked at $t = 0$. For $t > 0^+$, then we will have

---

Supported, in part, by the Office of Naval Research under Contract No. N00014-76-C-0326.

(2) $$m\frac{d^2y}{dt^2} + \beta\frac{dy}{dt} + ky = 0, \quad y(0^+) = 0, \quad \frac{dy}{dt}(0^+) = I_0/m.$$

(The last condition follows by integrating (1) from 0^- to 0^+ where the displacement remains zero.)

A regular perturbation problem would result if we sought an approximate solution on any finite t interval for (relatively) small values of the damping constant β (that is $\beta^2 \ll mk$). The weakly damped oscillator problem has a well-known solution (see Cole) which depends analytically on β and tends to the solution of the undamped oscillator problem as $\beta \to 0$. There is necessarily nonuniform convergence at $t = \infty$, however, since the undamped oscillator continues its motion with undiminished maximum amplitude, while the slightly damped solution ultimately decays to zero. A two-variable expansion procedure (see Cole or, for example, Greenlee and Snow [30]) is most appropriate to describe the asymptotic behavior near $t = \infty$, though we shall not pursue the matter.

A singular perturbation problem occurs when the mass m is relatively small. (If we ignored the mass, there would be a non-negligible effect on the solution.) Nondimensionalizing the initial value problem (2), let us set

$$\bar{t} = kt/\beta \quad \text{and} \quad \bar{y} = \beta y/I_0$$

to obtain the dimensionless problem

(3) $$\varepsilon\frac{d^2\bar{y}}{dt^2} + \frac{d\bar{y}}{dt} + \bar{y} = 0, \quad \bar{y}(0) = 0, \quad \frac{d\bar{y}}{dt}(0) = \frac{1}{\varepsilon},$$

with the small, positive parameter $\varepsilon = mk/\beta^2$ (that is, we shall suppose mk is small compared to β^2, so we simultaneously examine the problem with a relatively small spring constant k or large damping constant β. We note that mk/β^2 had an infinite limit (instead of zero) in the small damping problem).

Omitting the bars, the exact solution of the initial value problem (3) is given by

(4) $$y(t, \varepsilon) = \frac{1}{\varepsilon(\rho_1 - \rho_2)}\left(e^{\rho_1 t} - e^{\rho_2 t}\right)$$

where

$$\rho_1(\varepsilon) = [-1+\sqrt{1-4\varepsilon}]/2\varepsilon = -1 + O(\varepsilon)$$

and

$$\rho_2(\varepsilon) = [-1-\sqrt{1-4\varepsilon}]/2\varepsilon = -\frac{1}{\varepsilon} + 1 + O(\varepsilon)$$

are the roots of the characteristic polynomial $\varepsilon\rho^2 + \rho + 1 = 0$. (Here, the Landau symbol $O(\varepsilon)$ represents an error e such that $|e| \le k\varepsilon$ for some $k > 0$ and all sufficiently small $\varepsilon > 0$.) Since $\rho_1 \to -1$ and $\rho_2 \to -\infty$ as $\varepsilon \to 0$, we have

$$(5) \qquad y(t, \varepsilon) = e^{-t} - e^{-t/\varepsilon} + O(\varepsilon)$$

since, for example, $e^{-((1/\varepsilon)-1)t} = e^{-t/\varepsilon} + O(\varepsilon)$ for $t \ge 0$. (Note that the solution to the physically meaningless problem with $\varepsilon < 0$ would blow up as $\varepsilon \to 0$.) It is essential to note the nonuniform convergence of the solution which occurs at $t = 0$ as $\varepsilon \to 0^+$, that is, we have

$$\lim_{\varepsilon \to 0} y(t, \varepsilon) = \begin{cases} 0, & t = 0, \\ e^{-t}, & t > 0, \end{cases}$$

since $e^{-t/\varepsilon}$ is asymptotically negligible for $t > 0$. Such nonuniform convergence is the hallmark of singular perturbation problems. It is also important to realize that the limiting solution

$$(6) \qquad Y_0(t) = e^{-t}$$

for $t > 0$ satisfies the "reduced" equation $Y_0' + Y_0 = 0$, but neither of the initial conditions prescribed for y. Why Y_0 selects the initial value 1 is a mystery still to be explained.

Using the convergent expansions for the $\rho_i(\varepsilon)$'s as $\varepsilon \to 0$, the solution (4) of (3) is seen to be of the form

$$(7) \qquad y(t, \varepsilon) = Y(t, \varepsilon) + \Pi(\tau, \varepsilon)$$

where the "outer" solution

$$Y(t, \varepsilon) = e^{\rho_1 t}/\varepsilon(\rho_1 - \rho_2)$$

has an asymptotic expansion

$$(8) \qquad Y(t, \varepsilon) \sim \sum_{j=0}^{\infty} Y_j(t)\varepsilon^j$$

as $\varepsilon \to 0$, for all $t \ge 0$, usually called the outer expansion, and the "boundary layer" correction

$$\Pi(\tau, \varepsilon) = e^{\varepsilon\rho_2 \tau}/\varepsilon(\rho_1 - \rho_2)$$

in the "stretched" (or boundary layer) variable

$$(9) \qquad \tau = t/\varepsilon$$

has an asymptotic expansion

(10) $$\Pi(\tau, \varepsilon) \sim \sum_{j=0}^{\infty} \Pi_j(\tau)\varepsilon^j$$

whose terms Π_j all tend to zero as τ tends to infinity. (Olver [62] is an excellent reference concerning asymptotic expansions. We note that such expansions are usually divergent, rather than convergent, and that (8) should be interpreted as meaning that for any integer $N \geq 0$,

$$Y(t, \varepsilon) - \sum_{j=0}^{N} Y_j(t)\varepsilon^j = o(\varepsilon^N)$$

(that is, the right hand side tends to zero as $\varepsilon \to 0$ at a rate faster than ε^N).) The terms of the expansions (8) and (10) can be uniquely obtained by a Taylor series expansion of $Y(t, \varepsilon)$ and $\Pi(\tau, \varepsilon)$ about $\varepsilon = 0$ or, more directly, via the straightforward procedure given below.

Since the solution (7) is asymptotically equal to the outer solution (8) for $t > 0$ (that is, $\Pi \sim 0$), the asymptotic series for Y must satisfy

$$\varepsilon Y'' + Y' + Y = 0$$

as a power series in ε. Equating coefficients of ε^j for each $j \geq 0$ requires that $Y_j' + Y_j = -Y_{j-1}''$, so

(11) $$Y_j(t) = e^{-t}Y_j(0) - \int_0^t e^{-(t-s)} Y_{j-1}''(s)ds, \quad j \geq 0,$$

with $Y_{-1}(t) \equiv 0$. Thus, the outer expansion (8) will be completely and uniquely obtained termwise up to specification of the initial value

$$Y(0, \varepsilon) \sim \sum_{j=0}^{\infty} Y_j(0)\varepsilon^j.$$

Since y and Y both satisfy the differential equation of (3) and $\dfrac{d^k}{dt^k} = \dfrac{1}{\varepsilon^k} \dfrac{d^k}{d\tau^k}$, the boundary layer correction Π must satisfy

$$\frac{d^2\Pi}{d\tau^2} + \frac{d\Pi}{d\tau} + \varepsilon\Pi = 0, \quad \tau \geq 0,$$

as a power series in ε. Thus, $d^2\Pi_j/d\tau^2 + d\Pi_j/d\tau = -\Pi_{j-1}$ for each $j \geq 0$. Asking that $\Pi_j \to 0$ as $\tau \to \infty$, we then have

(12) $$\Pi_j(\tau) = -e^{-\tau}\frac{d\Pi_j(0)}{d\tau} - \int_0^\tau \int_0^r e^{-(\tau-r)}\Pi_{j-1}(r)dr$$

so the boundary layer (correction's) expansion (10) is determined termwise up to selection of its initial derivative

$$\frac{d\Pi}{d\tau}(0, \varepsilon) \sim \sum_{j=0}^{\infty} \frac{d\Pi_j(0)}{d\tau} \varepsilon^j = - \sum_{j=0}^{\infty} \Pi_j(0)\varepsilon^j .$$

The representation (7) implies the "matching" condition

$$y'(0, \varepsilon) = \frac{1}{\varepsilon} = Y'(0, \varepsilon) + \frac{1}{\varepsilon} \frac{d\Pi}{d\tau}(0, \varepsilon) .$$

Hence, we must have

(13)
$$\begin{cases} \dfrac{d\Pi_0(0)}{d\tau} = 1 , \\ \dfrac{d\Pi_j(0)}{d\tau} = -Y'_{j-1}(0) \quad \text{for each } j \geq 1 . \end{cases}$$

Thus, the initial values needed for the boundary layer correction terms are determined from earlier terms of the outer solution. In particular, $\Pi_0(\tau)$ is now completely known. Likewise, the remaining initial condition

$$y(0, \varepsilon) = 0 = Y(0, \varepsilon) + \Pi(0, \varepsilon)$$

implies that we must have

(14)
$$Y_j(0) = -\Pi_j(0) \quad \text{for each } j \geq 0 .$$

Thus Π_0 determines Y_0 and, more generally, the terms Y_j of the outer expansion can be determined in a termwise bootstrap fashion, since $\Pi_j(0)$ depends only on $Y'_{j-1}(0)$. Indeed, (12) and (13) imply that

$$Y_j(0) = -Y'_{j-1}(0) \quad \text{for each } j \geq 1 .$$

Our formal procedure, then, produces the asymptotic solution

(15)
$$y(t, \varepsilon) = \left(e^{-t} - e^{-\tau}\right) + \varepsilon\{e^{-t}(t-1) - e^{-\tau}(\tau-1)\} + O(\varepsilon^2)$$

in agreement with the exact solution (4). This result for $\tau = t/\varepsilon$ clearly displays the rapid initial rise in displacement obtained for small ε, followed by an ultimate decay like a massless system. We note that the boundary layer calculation, leading to the representation of terms (12), played an essential role in obtaining the asymptotic solution $Y(t, \varepsilon)$ appropriate for $t > 0$. In particular, knowing that $\Pi_0(\tau) = -e^{-\tau}$ implied that the maximum displacement of the system tends to one as $\varepsilon \to 0$. Pictorially, we have the displacements

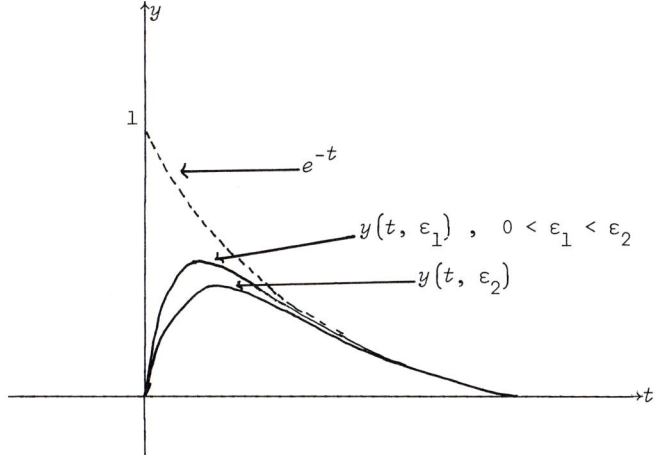

We note that Chapter 1 of Andronov, Vitt, and Khaikin [2] considers the oscillator with small mass somewhat more intuitively, developing the idea of an initial jump.

For the corresponding two-point problem

(16) $\quad \varepsilon y'' + y' + y = 0, \quad y(0) = 0, \quad y(1) = e^{-1},$

the unique solution is again given by

$$y(t, \varepsilon) = Y_0(t) + \Pi_0(\tau) + O(\varepsilon)$$

with Y_0 and Π_0 as before. The limiting solution $Y_0(t)$ for $t > 0$, however, now satisfies the reduced problem

$$Y_0' + Y_0 = 0, \quad Y_0(1) = e^{-1}$$

obtained by using the differential equation (16) with $\varepsilon = 0$ and cancelling the initial condition. Likewise, for the initial value problem

(17) $\quad \varepsilon y'' + y' + y = 0, \quad y(0) = 0, \quad y'(0) = 1,$

the limiting solution Y_0 for $t > 0$ will be the trivial solution of the reduced problem

$$Y_0' + Y_0 = 0, \quad Y_0(0) = 0.$$

Cancellation of a boundary condition to define the reduced problem is natural, since the differential equation is of first order when $\varepsilon = 0$. That no simple cancellation occurs in the solution of our oscillator problem (3) is because the boundary condition $y'(0) = 1/\varepsilon$ becomes singular as $\varepsilon \to 0^+$ (see O'Malley and Keller [72], however, for the definition of an appropriate cancellation rule). Boundary value problems for scalar differential equations with small parameters multiplying the highest derivatives are one of the best studied singular perturbation problems (see Wasow [86] or O'Malley [63]). Such problems and their generalizations do occur in control.

2. Simplified Models in Control and Systems Theory

Consider a physical system described by the equations

(1)
$$\begin{cases} \dot{x} = f(x, y, z, u, t), \\ \varepsilon \dot{y} = g(x, y, z, u, t), \\ \dot{z} = \frac{1}{\mu} h(x, y, z, u, t), \end{cases}$$

where x, y, z, and u are vectors, ε is a small positive parameter, and μ is a large positive parameter. Roughly, y corresponds to a fast-varying vector and z to a slowly-varying vector (compared to x). It would be natural to attempt to simplify the system by neglecting the small parameters ε and $1/\mu$ and solving the reduced system

(2)
$$\begin{cases} \dot{X} = f(X, Y, Z, u, t), \\ 0 = g(X, Y, Z, u, t), \\ \dot{Z} = 0. \end{cases}$$

Then we would have

(3)
$$\begin{cases} Z = \text{constant} \\ \text{and} \\ Y = \phi(X, Z, u, t) \end{cases}$$

presuming that we could find a unique root of the nonlinear equation

$$g(X, \phi, Z, u, t) = 0.$$

Thus we would be left with the lower-dimensional, non-"stiff" model

(4)
$$\dot{X} = f\bigl(X, \phi(X, Z, u, t), Z, u, t\bigr) = F(X, Z, u, t).$$

Such approximations are common in many areas of science, for example, an analogous procedure is known as the prompt jump approximation in nuclear reactor theory (see Hetrick [35]) and as the pseudo-steady state hypothesis in enzyme kinetics (see Rubinow [78]) and is basic to the development of numerical methods for integrating stiff differential equations (see Willoughby [88]). We need to determine to what extent such simplifications are valid.

Desoer [21], [22], uses ε to indicate the degree of smallness of certain "stray" elements (for example, stray capacitances and lead inductances) in circuits, and he uses μ to represent "sluggish" elements (like chokes and coupling capacitors). The stray elements will affect the high-frequency behavior, while the sluggish elements affect the low-frequency behavior. On finite t-intervals, we have a singular perturbation of (2) by including the ε terms of (1) and a regular perturbation by including the $1/\mu$ terms. Stability considerations for appropriate high-frequency ("boundary layer") models will be needed to justify the mid-frequency (or reduced) model (4). We shall not discuss the appropriate low-frequency approximations, noting only that they must deal with nonuniform convergence at $t = \infty$ and a regular perturbation analysis for finite t. Many other circuit theory examples are given in Andronov, Vitt, and Khaikin [2]. Sophisticated discussions of regular perturbation theory include Rellich [75] and Kato [46].

Since, in practice, one always neglects some small parasitics, Kokotović has claimed that all control problems are singularly perturbed. Successful control engineers must, then, naturally use their intuition to check the hypotheses of the theorems which guarantee legitimacy to the reduced order models which they use.

Sannuti and Kokotović [80] gave an example of a voltage regulator described by a linear system

$$\begin{cases} \dfrac{dx}{dt} = a_1 x + a_2 z, \\[2mm] \varepsilon \dfrac{dz}{dt} = a_3 z + bu, \end{cases}$$

for $t \geq 0$ with

$$a_1 = 0.1 \begin{pmatrix} -2 & 5 \\ 0 & -5 \end{pmatrix}, \quad a_2 = \begin{pmatrix} 0 & 0 & 0 \\ 1.6 & 0 & 0 \end{pmatrix},$$

$$a_3 = \begin{pmatrix} -10/7 & 60/7 & 0 \\ 0 & -2.5 & -7.5 \\ 0 & 0 & -1 \end{pmatrix}, \quad \text{and} \quad b = \begin{pmatrix} 0 \\ 0 \\ 3 \end{pmatrix},$$

for the not-so-small parameter $\varepsilon = 0.1$. Here, the object is to minimize the cost functional

$$J = \tfrac{1}{2} \int_0^\infty \left[x' \begin{pmatrix} 1 & 0 \\ 0 & 0 \end{pmatrix} x + u^2 \right] dt .$$

Setting $\varepsilon = 0$ corresponds to neglecting the small time constants. Even though $\varepsilon = 0.1$ is not very small, the second-order reduced problem provides an acceptable solution, which is much easier to compute than the exact solution of the full fifth-order model. (Since a_1 is also small, one might simultaneously also neglect it.) Sannuti and Kokotović observe that it does not work to integrate the full system for $\varepsilon = 0.1$ with the feedback solution appropriate for $\varepsilon = 0$, but one should not expect it to. Using an improved feedback approximation, however, the difficulty disappears.

Likewise, Kokotović and Yackel [49] discuss a model for speed control of a small dc motor described by the state equations

$$\begin{cases} \dfrac{d\omega}{dt} = Di/G\ , \\ \varepsilon L \dfrac{di}{dt} = -C\omega - R_a i + v\ . \end{cases}$$

Since the armature inductance εL is typically small, it is common to use the simplified model

$$\frac{d\Omega}{dt} = \frac{D}{GR_a}(-C\Omega + v)$$

in designing servosystems. This acknowledges the fact that mechanical time constants are large compared to electrical ones, but the model will not be appropriate for a fast initial transient. Finally, more examples are found in Chow and Kokotović [14] and elsewhere throughout the literature.

3. The Singularly Perturbed Linear State Regulator Problem

A. PROBLEM FORMULATION

Consider the problem of minimizing the scalar cost functional

(1) $\quad J = \tfrac{1}{2} y'(1,\varepsilon)\Pi(\varepsilon)y(1,\varepsilon) + \tfrac{1}{2}\displaystyle\int_0^1 [y'(t,\varepsilon)Q(t,\varepsilon)y(t,\varepsilon) + u'(t,\varepsilon)u(t,\varepsilon)]dt$

where the vector $y = \begin{pmatrix} x \\ z \end{pmatrix}$ satisfies the singularly perturbed system of state equations

(2) $\quad \begin{cases} \dfrac{dx}{dt} = A_1(t,\varepsilon)x + A_2(t,\varepsilon)z + B_1(t,\varepsilon)u\ , \\ \varepsilon \dfrac{dz}{dt} = A_3(t,\varepsilon)x + A_4(t,\varepsilon)z + B_2(t,\varepsilon)u\ , \end{cases}$

on $0 \le t \le 1$ with prescribed initial vectors

(3) $\quad\quad\quad\quad\quad\quad x(0,\varepsilon)\ \text{and}\ z(0,\varepsilon)\ .$

For symmetric, positive semi-definite matrices Π and Q and a fixed $\varepsilon > 0$, such problems have a well-known unique solution (see Kalman [44]). We would like to determine the limiting behavior of this solution as the small positive parameter ε tends towards zero.

We shall not discuss several important linear generalizations of this problem which have already been dealt with in the control literature. They include the fixed endpoint problem (see Wilde and Kokotović [87]), several parameter problems (see Asatani [3]), distributed parameter systems (see Lions [57]), stochastic problems (see, for example, Blankenship and Sachs [5] and Haddad [32]), and systems with small time-delays (see Sannuti and Reddy [81]). We will, however, consider a nonlinear generalization.

Let us take states x, z, and the control u to be vectors of dimensions m, n, and r, respectively; assume that the matrices A_i, B_i, Q, and Π and the vectors $x(0, \varepsilon)$ and $z(0, \varepsilon)$ all have asymptotic power series expansions as $\varepsilon \to 0$; that the A_i, B_i, and Q are infinitely differentiable functions of t; and that Π has the partitioned form

(4) $$\Pi(\varepsilon) = \begin{pmatrix} \Pi_1(\varepsilon) & \varepsilon\Pi_2(\varepsilon) \\ \varepsilon\Pi_2'(\varepsilon) & \varepsilon\Pi_3(\varepsilon) \end{pmatrix}$$

with blocks having sizes compatible with the dimensions of x and z. We note, in particular, that (4) implies that the terminal cost term

(5) $$\lambda(\varepsilon) \equiv \tfrac{1}{2} y'(1, \varepsilon)\Pi(\varepsilon)y(1, \varepsilon)$$

of (1) depends only on the "slow" state x when $\varepsilon = 0$. We could considerably curtail the hypotheses specifying smoothness with respect to t and ε with little loss, but the nature of the asymptotic solution would be different if the restriction (4) were not assumed (see Glizer and Dmitriev [28]).

Following Kalman [44] and, for example, Anderson and Moore [1] or Coppel [20]), we introduce the Hamiltonian

(6) $$h(x, z, p, q, u, t, \varepsilon) = (x'z')Q(t, \varepsilon)\binom{x}{z} + u'u + p'(A_1 x + A_2 z + B_1 u) + q'(A_3 x + A_4 z + B_2 u)$$

for m and n dimensional costate vectors p and εq which satisfy the linear adjoint (or costate) equations

(7) $$\frac{dp}{dt} = -\frac{\partial h}{\partial x} \quad \text{and} \quad \varepsilon \frac{dq}{dt} = -\frac{\partial h}{\partial z}$$

and the terminal conditions

(8) $$p(1, \varepsilon) = \frac{\partial \lambda(\varepsilon)}{\partial x(1,\varepsilon)} \quad \text{and} \quad \varepsilon q(1, \varepsilon) = \frac{\partial \lambda(\varepsilon)}{\partial z(1,\varepsilon)}$$

for λ defined by (5). (Here we have used the costate εq to compensate for the fact that dz/dt is formally of order $O(1/\varepsilon)$. We also note that the state equations (2) take the canonical form $\frac{dx}{dt} = \frac{\partial h}{\partial p}$ and $\varepsilon \frac{dz}{dt} = \frac{\partial h}{\partial q}$.) As Kalman showed, minimization of the cost functional is equivalent to minimizing the Hamiltonian. Thus, setting

$$\frac{\partial h}{\partial u} = u + B_1'p + B_2'q = 0$$

provides the unique minimum since $\partial^2 h/\partial u^2 = I_r$ is positive definite. The optimal control is therefore given by

(9) $$u = -B_1'p - B_2'q ,$$

so eliminating u in (2) leaves us with a linear singularly perturbed two-point boundary value problem for the states x and z and the (scaled) costates p and q.

Our linear-quadratic regulator problem has thereby been reduced to analyzing the asymptotic behavior of the $2m + 2n$ dimensional linear system

(10)
$$\begin{cases} \dfrac{dx}{dt} = A_1 x - B_1 B_1' p - A_2 z - B_1 B_2' q , & x(0, \varepsilon) \text{ prescribed,} \\[4pt] \dfrac{dp}{dt} = -Q_1 x - A_1' p - Q_2 z - A_3' q , & p(1, \varepsilon) = \Pi_1(\varepsilon) x(1, \varepsilon) + \varepsilon \Pi_2(\varepsilon) z(1, \varepsilon) , \\[4pt] \varepsilon \dfrac{dz}{dt} = A_3 x - B_2 B_1' p + A_4 z - B_2 B_2' q , & z(0, \varepsilon) \text{ prescribed,} \\[4pt] \varepsilon \dfrac{dq}{dt} = -Q_2' x - A_2' p - Q_3 z - A_4' q , & q(1, \varepsilon) = \Pi_2'(\varepsilon) x(1, \varepsilon) + \Pi_3(\varepsilon) z(1, \varepsilon) , \end{cases}$$

where $Q = \begin{pmatrix} Q_1 & Q_2 \\ Q_2' & Q_3 \end{pmatrix}$.

Linear singular perturbation problems such as (10) have been well studied. We note, for example, that Harris [34] considers linear boundary value problems for the $p + q$ dimensional system

(11)
$$\begin{cases} u' = A(t, \varepsilon)u + B(t, \varepsilon)v , \\[4pt] \varepsilon v' = C(t, \varepsilon)u + D(t, \varepsilon)v , \end{cases}$$

on $0 \le t \le 1$ under the principal assumption that the eigenvalues of the $q \times q$ matrix $D(t, 0)$ have nonzero real parts throughout $[0, 1]$. He shows that such systems have a p dimensional manifold of solutions which tend to solutions of the reduced system

$$\begin{cases} U_0' = A(t, 0)U_0 + B(t, 0)V_0 , \\[4pt] 0 = C(t, 0)U_0 + D(t, 0)V_0 , \end{cases}$$

as $\varepsilon \to 0$. That system has p linearly independent solutions determined by

$$U_0' = \left(A(t, 0) - B(t, 0)D^{-1}(t, 0)C(t, 0) \right) U_0$$

since $V_0 = -D^{-1}(t, 0)C(t, 0)U_0$. Moreover, if $D(t, 0)$ has k, $0 \le k \le q$, stable eigenvalues, there are k linearly independent solutions of (11) which decay to zero

$\left(\text{like } e^{-C_1 t/\varepsilon} \text{ for some positive definite matrix } C_1\right)$ as the stretched variable

(12) $$\tau = t/\varepsilon$$

tends to infinity, and there will be $q - k$ linearly independent solutions which decay to zero $\left(\text{like } e^{-C_2(1-t)/\varepsilon} \text{ for some } C_2 > 0\right)$ as

(13) $$\sigma = (1-t)/\varepsilon$$

tends to infinity. (In a sense, this theory produces a fundamental matrix for (11) which is asymptotically valid as $\varepsilon \to 0$ in $0 \le t \le 1$ (see Turrittin [82]). The results are proved by integral equation techniques.) Since the general solution of (11) is a linear combination of any $p + q$ linearly independent asymptotic solutions, the behavior of any solution to (11) satisfies a three time-scale property, that is, such an asymptotic solution must be an additive sum of functions depending on t, τ, and σ, respectively, with the function of τ (or σ) providing boundary layer behavior (that is, nonuniform convergence as $\varepsilon \to 0$) at $t = 0$ (or $t = 1$) and the limiting solution within $(0, 1)$ being a function of t which satisfies the reduced system there. Much more complicated behavior would result if we allowed the eigenvalues of D to cross the imaginary axis or to remain on it (an exchange of stability or neutral stability).

In the problem (10), the role of $D(t, 0)$ is played by the $2n \times 2n$ Hamiltonian matrix

(14) $$G(t) = \begin{pmatrix} A_4(t,0) & -B_2(t,0)B_2'(t,0) \\ -Q_3(t,0) & -A_4'(t,0) \end{pmatrix}.$$

Because $J_n G = -G' J_n$ is symmetric for the symplectic matrix

(15) $$J_n = \begin{pmatrix} 0 & I_n \\ -I_n & 0 \end{pmatrix},$$

corresponding to any eigenvalue λ of G is another eigenvalue $-\lambda$. Thus, G will be invertible and the results corresponding to those for (11) will hold provided

(H1) *All eigenvalues of the matrix* $G(t)$ *have nonzero real parts throughout* $0 \le t \le 1$.

Indeed, G will then have n stable eigenvalues and n unstable ones, so the singularly perturbed system (10) will have n linearly independent solutions which decay to zero away from $t = 0$, n others which decay to zero for $t < 1$, and $2m$ others which satisfy the reduced system corresponding to (10) in the limit $\varepsilon \to 0$. We note that (H1) relates to the factorizability of a related characteristic polynomial and to stabilizability of associated control problems (see Coppel [19]).

B. THE REDUCED PROBLEM

The reduced system for (10) has the form

(16)
$$\begin{cases} \dfrac{d}{dt}\begin{pmatrix} X_0 \\ P_{10} \end{pmatrix} = M \begin{pmatrix} X_0 \\ P_{10} \end{pmatrix} + J_n L' J_n \begin{pmatrix} Z_0 \\ P_{20} \end{pmatrix}, \\ \\ 0 = L \begin{pmatrix} X_0 \\ P_{10} \end{pmatrix} + G \begin{pmatrix} Z_0 \\ P_{20} \end{pmatrix}, \end{cases}$$

for

$$M = \begin{pmatrix} A_{10} & -B_{10}B'_{10} \\ -Q_{10} & -A'_{10} \end{pmatrix} \quad \text{and} \quad L = \begin{pmatrix} A_{30} & -B_{20}B'_{10} \\ -Q'_{20} & -A'_{20} \end{pmatrix},$$

where, for example, $A_{10} = A_1(t, 0)$. (Here we have used P_{10} and P_{20}, instead of P_0 and Q_0, to represent costates, to avoid confusion between Q_0 and submatrices or expansion coefficients of Q.) It is natural to retain the limiting boundary conditions

(17) $\qquad X_0(0) = x(0, 0) \quad \text{and} \quad P_{10}(1) = \Pi_{10}(0) X_{10}(1)$

of (10) for (16), thereby defining a reduced boundary value problem. Then, however, the reduced problem (16)-(17) cannot be expected to provide the limiting solution to (10) near $t = 0$ or $t = 1$ since it fails to account for the initial condition for z or the terminal condition for q in terms of x and z. Its tremendous advantage over (10), however, is having differential order $2m$ instead of $2m + 2n$.

Since (H1) allows us to obtain $\begin{pmatrix} Z_0 \\ P_{20} \end{pmatrix}$ as a linear function of $\begin{pmatrix} X_0 \\ P_{10} \end{pmatrix}$, (16)-(17) is equivalent to the boundary value problem

(18) $\qquad \dfrac{d}{dt}\begin{pmatrix} X_0 \\ P_{10} \end{pmatrix} = V(t) \begin{pmatrix} X_0 \\ P_{10} \end{pmatrix}, \quad X_0(0) = x(0, 0), \quad P_{10}(1) = \Pi_1(0) X_0(1)$

for $V = M - J_m L' J_n G^{-1} L$. Further, the Hamiltonian structure of M and G^{-1} implies that of V, that is, $J_m V = -V' J_m = J_m M + L' J_n G^{-1} L$ is symmetric, so that (18) becomes

(19) $\qquad \dfrac{d}{dt}\begin{pmatrix} X_0 \\ P_{10} \end{pmatrix} = \begin{pmatrix} V_1 & V_2 \\ V_3 & -V'_1 \end{pmatrix}\begin{pmatrix} X_0 \\ P_{10} \end{pmatrix}, \quad X_0(0) = x(0, 0), \quad P_{10}(1) = \Pi_1(0) X_0(1)$

with

$$V_1 = A_{10} + \begin{pmatrix} B_{10}B'_{20} & A_{20} \end{pmatrix} J_n G^{-1} \begin{pmatrix} A_{30} \\ -Q'_{20} \end{pmatrix},$$

$$V_2 = -B_{10}B'_{10} - \begin{pmatrix} B_{10}B'_{20} & A_{20} \end{pmatrix} J_n G^{-1} \begin{pmatrix} B_{20}B'_{10} \\ A'_{20} \end{pmatrix} = V'_2,$$

and

$$V_3 = -Q_{10} + \begin{pmatrix} A'_{30} & -Q_{20} \end{pmatrix} J_n G^{-1} \begin{pmatrix} A_{30} \\ -Q'_{20} \end{pmatrix} = V'_3.$$

Thus the reduced problem (19) is an mth order regulator problem, which we shall call the reduced regulator problem, and it is natural to seek a solution to it in the feedback form

(20) $$P_{10} = K_{10} X_0$$

where the symmetric $m \times m$ matrix K_{10} satisfies the matrix Riccati differential equation

(21) $$\dot{K}_{10} = -K_{10}V_1 - V'_1 K_{10} - K_{10}V_2 K_{10} + V_3, \quad K_{10}(1) = \Pi_1(0)$$

(see again Kalman [44]). If K_{10} exists back to $t = 0$, we only need to integrate the initial value problem

(22) $$\dot{X}_0 = (V_1 + V_2 K_{10}) X_0, \quad X_0(0) = x(0, 0)$$

to completely solve the reduced problem (16)-(17).

According to Bucy [6], necessary and sufficient conditions to solve the linear Hamiltonian system (19) are

(H2) *The* $m \times m$ *matrices* $V_2(t)$ *and* $V_3(t)$ *are both negative semi-definite throughout* $0 \leq t \leq 1$.

We conjecture that (H2) is redundant. To actually calculate V_2 and V_3, we would have to obtain the blocks of G^{-1} (for a method to do so, see Theorem 5 of Coppel [19] and the calculations of O'Malley and Kung [74]). For A_{40} invertible, O'Malley and Kung showed that $V_2 \leq 0$ while L. Anderson (personal communication) has since shown that $V_3 \leq 0$ then holds. Likewise, O'Malley [65] found that (H2) held when x, z, and u are scalars.

Our Riccati solution of the reduced regulator problem suggests that the original problem could also be solved through a Riccati feedback approach and that is true (see

Yackel and Kokotović [91] and O'Malley and Kung [73]). One would set $\binom{p}{q} = k(t, \varepsilon)\binom{x}{z}$. That approach is nontrivial (but also important in other contexts) because the Riccati equation for k is singularly perturbed, that is, it has a small parameter multiplying its derivative term.

Note that another (perhaps more natural) reduced problem is obtained by setting $\varepsilon = 0$ in the original optimal control problem (1)-(4). Thus, suppose we

(23)
$$\begin{cases} \text{minimize } J = \tfrac{1}{2}X_0'(1)\Pi_1(0)X_0(1) + \tfrac{1}{2}\int_0^1 \left\{ \begin{pmatrix} X_0' & Z_0' \end{pmatrix} Q_0(t) \begin{pmatrix} X_0 \\ Z_0 \end{pmatrix} + U_0'U_0 \right\} dt \text{ with} \\ \dot{X}_0 = A_{10}X_0 + A_{20}Z_0 + B_{10}U_0, \quad X_0(0) = x(0, 0), \\ 0 = A_{30}X_0 + A_{40}Z_0 + B_{20}U_0. \end{cases}$$

Here both Z_0 and U_0 play the role of control variables, while X_0 remains a state vector. If A_{40}^{-1} exists, we can find Z_0 as a linear function of X_0 and U_0 and (23) reduces to a standard linear regulator problem in the form

(24)
$$\begin{cases} \text{minimize } J = \tfrac{1}{2}X_0'(1)\Pi_1(0)X_0(1) + \tfrac{1}{2}\int_0^1 \{-X_0'(t)V_3(t)X_0(t) + W_0'(t)R(t)W_0(t)\} dt \\ \text{with} \quad \dot{X}_0 = V_1(t)X_0 + B(t)W_0, \quad X_0(0) = x(0, 0). \end{cases}$$

Here W_0 is a linear combination of U_0 and X_0; V_1, V_2, and V_3 are the submatrices of (19); and R is a positive-definite matrix such that $V_2 = -BR^{-1}B'$. The equivalence of the reduced regulator problems (19) and (24) follows under the hypotheses of O'Malley and Kung [74], but we expect it to be generally true under hypothesis (H1).

C. BOUNDARY LAYERS

Since the matrix $G(t)$ has n stable eigenvalues and n unstable ones, the general theory for linear singularly perturbed boundary value problems and some experience suggest that we seek an asymptotic solution to our two-point problem (10) in the form

(25)
$$\begin{cases} x(t, \varepsilon) = X(t, \varepsilon) + \varepsilon m_1(\tau, \varepsilon) + \varepsilon n_1(\sigma, \varepsilon), \\ z(t, \varepsilon) = Z(t, \varepsilon) + m_2(\tau, \varepsilon) + n_2(\sigma, \varepsilon), \\ p(t, \varepsilon) = P_1(t, \varepsilon) + \varepsilon \rho_1(\tau, \varepsilon) + \varepsilon \gamma_1(\sigma, \varepsilon), \\ q(t, \varepsilon) = P_2(t, \varepsilon) + \rho_2(\tau, \varepsilon) + \gamma_2(\sigma, \varepsilon), \end{cases}$$

where the outer expansion

(26) $$(X, Z, P_1, P_2) \sim \sum_{j=0}^{\infty} (X_j, Z_j, P_{1j}, P_{2j}) \varepsilon^j$$

provides the asymptotic solution to (10) within (0, 1) ; the initial boundary layer correction

(27) $$(\varepsilon m_1, m_2, \varepsilon \rho_1, \rho_2) \sim \sum_{j=0}^{\infty} (\varepsilon m_{1j}, m_{2j}, \varepsilon \rho_{1j}, \rho_{2j}) \varepsilon^j$$

satisfies (10) and its terms tend to zero as the stretched variable

$$\tau = t/\varepsilon$$

tends to infinity; and the terms of the terminal boundary layer correction

(28) $$(\varepsilon n_1, n_2, \varepsilon \gamma_1, \gamma_2) \sim \sum_{j=0}^{\infty} (\varepsilon n_{1j}, n_{2j}, \varepsilon \gamma_{1j}, \gamma_{2j}) \varepsilon^j$$

tend to zero as

$$\sigma = (1-t)/\varepsilon$$

tends to infinity. In part, we write these forms of the asymptotic solution to display its three time scale structure and the relative importance of the different scales to the different components of the asymptotic solution. In practice, one would typically compute only the ε^0 and ε^1 coefficients. We further note that the control relation (9) and the representation (25) imply that the optimal control will have a corresponding asymptotic representation

(29) $$u(t, \varepsilon) = U(t, \varepsilon) + v(\tau, \varepsilon) + w(\sigma, \varepsilon)$$

where, for example,

$$v(\tau, \varepsilon) = -\varepsilon B_1'(\varepsilon\tau, \varepsilon)\rho_1(\tau, \varepsilon) - B_2'(\varepsilon\tau, \varepsilon)\rho_2(\tau, \varepsilon)$$

and the boundary layer corrections v and w are asymptotically significant only near the endpoints $t = 0$ and $t = 1$, respectively. Since u has the form

$$u(t, \varepsilon) = U_0(t) + v_0(\tau) + w_0(\sigma) + O(\varepsilon)$$

the optimal control will generally converge nonuniformly near each endpoint and a boundary layer analysis is necessary to determine the endpoint control. A typical plot of optimal control is pictured in the figure.

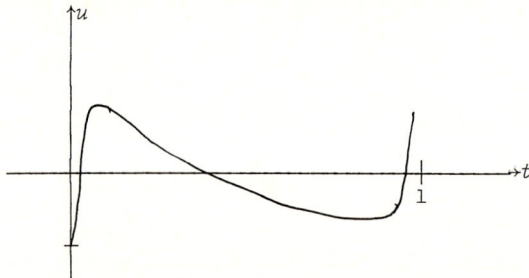

Finally, the expansions (25) and (29) imply that optimal cost will have the form

(30) $$J^*(\varepsilon) \sim \sum_{k=0}^{\infty} J_k \varepsilon^k$$

where the leading term J_0 is the optimal cost for the reduced regulator problem (19), that is,

(31) $$J_0 = \tfrac{1}{2} x'(0, 0) K_{10}(0) x(0, 0)$$

(see (20) and Kalman [44]). The boundary layer contributions to the cost, like the integral $\int_0^1 e^{-t/\varepsilon} dt$, are $O(\varepsilon)$.

We must now learn how to calculate the asymptotic solution (25). Since the boundary layer correction terms become negligible within $(0, 1)$, the outer expansion (26) must satisfy the differential system of (10) as a power series in ε. The leading terms $(X_0, Z_0, P_{10}, P_{20})$ will necessarily satisfy the limiting system (16) and, by the form of (25), the boundary conditions (17). (Unlike the spring-mass system, then, the boundary conditions appropriate for the limiting solution here are obtained without first calculating a boundary layer term.) Under hypotheses (H1) and (H2), the resulting reduced problem (16)-(17) has a unique solution. Higher order terms in (26) will then satisfy nonhomogeneous forms of (16)-(17) with successively known forcing terms. The Fredholm alternative, then, guarantees that they, too, will have unique solutions.

Since the outer solution (26) accounts termwise for the initial condition for x and the terminal condition for p, the initial boundary layer correction (27) must adjust for any "boundary layer jump" $z(0, \varepsilon) - Z(0, \varepsilon)$ while the terminal correction (28) must account for the terminal condition for q. (We recall that Z_0 and P_{20} were determined from algebraic equations.) Since the solution of (10) will be asymptotically the sum of the outer expansion (26) and the initial boundary layer correction (27) near $t = 0$ (σ being asymptotically infinite), while (26) satisfies (10), it follows that (27) must satisfy (10) as a function of τ. Thus we seek a decaying solution of the linear system

(32)
$$\begin{cases}
\frac{dm_1}{d\tau} = \varepsilon A_1(\varepsilon\tau, \varepsilon)m_1 + A_2(\varepsilon\tau, \varepsilon)m_2 - \varepsilon B_1(\varepsilon\tau, \varepsilon)B_1'(\varepsilon\tau, \varepsilon)\rho_1 \\
\qquad\qquad\qquad\qquad\qquad\qquad\qquad - B_1(\varepsilon\tau, \varepsilon)B_2'(\varepsilon\tau, \varepsilon)\rho_2 , \\
\frac{d\rho_1}{d\tau} = \varepsilon Q_1(\varepsilon\tau, \varepsilon)m_1 - Q_2(\varepsilon\tau, \varepsilon)m_2 - \varepsilon A_1'(\varepsilon\tau, \varepsilon)\rho_1 - A_3'(\varepsilon\tau, \varepsilon)\rho_2 , \\
\frac{dm_2}{d\tau} = \varepsilon A_3(\varepsilon\tau, \varepsilon)m_1 + A_4(\varepsilon\tau, \varepsilon)m_2 - \varepsilon B_2(\varepsilon\tau, \varepsilon)B_1'(\varepsilon\tau, \varepsilon)\rho_1 \\
\qquad\qquad\qquad\qquad\qquad\qquad\qquad - B_2(\varepsilon\tau, \varepsilon)B_2'(\varepsilon\tau, \varepsilon)\rho_2 , \\
\frac{d\rho_2}{d\tau} = -\varepsilon Q_2(\varepsilon\tau, \varepsilon)m_1 - Q_3(\varepsilon\tau, \varepsilon)m_2 - \varepsilon A_2'(\varepsilon\tau, \varepsilon)\rho_1 - A_4'(\varepsilon\tau, \varepsilon)\rho_2 ,
\end{cases}$$

for $\tau \geq 0$ satisfying the initial condition

(33)
$$m_2(0, \varepsilon) \sim z(0, \varepsilon) - Z(0, \varepsilon) .$$

In particular, then, for $\varepsilon = 0$ we have the limiting constant coefficient boundary layer problem

(34)
$$\begin{cases}
\frac{dm_{10}}{d\tau} = A_{20}(0)m_{20} - B_{10}(0)B_{20}'(0)\rho_{20} , \\
\frac{d\rho_{10}}{d\tau} = -Q_{20}(0)m_{20} - A_{30}'(0)\rho_{20} , \\
\frac{dm_{20}}{d\tau} = A_{40}(0)m_{20} - B_{20}(0)B_{20}'(0)\rho_{20} , \quad m_{20}(0) = z(0, 0) - Z_0(0) , \\
\frac{d\rho_{20}}{d\tau} = -Q_{30}(0)m_{20} - A_{40}'(0)\rho_{20} .
\end{cases}$$

Presuming, then, that we can find an exponentially decaying solution to the initial value problem

(35)
$$\frac{d}{d\tau}\begin{pmatrix} m_{20} \\ \rho_{20} \end{pmatrix} = G(0)\begin{pmatrix} m_{20} \\ \rho_{20} \end{pmatrix} , \quad m_{20}(0) = z(0, 0) - Z_0(0)$$

we will determine the remaining decaying terms as

(36)
$$(m_{10}(\tau), \rho_{10}(\tau)) = -\int_\tau^\infty \left(\frac{dm_{10}(s)}{d\tau}, \frac{d\rho_{10}(s)}{d\tau}\right) ds$$

since $dm_{10}/d\tau$ and $d\rho_{10}/d\tau$ are linear combinations of m_{20} and ρ_{20}. We recall that $G(0)$ has half its eigenvalues stable and half unstable. Thus the decaying solutions of (35) are spanned by n linearly independent quasipolynomial solutions of the form

(37) $$s_i(\tau)e^{\lambda_{i0}\tau}, \quad i = 1, 2, \ldots, n,$$

where the s_i's are polynomials in τ and the λ_{i0}'s are stable eigenvalues of $G(0)$ (see, for example, Coddington and Levinson [17]), that is, we must have

$$\begin{pmatrix} m_{20} \\ \rho_{20} \end{pmatrix} = \sum_{i=1}^{n} s_i(\tau)e^{\lambda_{i0}\tau} k_i$$

for n appropriate vectors k_i. Let us assume

(H3a) *The* $n \times n$ *matrix* $T_{10} \equiv \begin{pmatrix} s_{11}(0) & s_{21}(0) & \ldots & s_{n1}(0) \end{pmatrix}$ *is nonsingular where the* n-*vectors* $s_{j1}(0)$ *are such that*

$$s_j(0) = \begin{pmatrix} s_{j1}(0) \\ s_{j2}(0) \end{pmatrix}, \quad j = 1, \ldots, n.$$

Then the solution of (35) *is uniquely given by*

(38) $$\begin{pmatrix} m_{20} \\ \rho_{20} \end{pmatrix} = \begin{pmatrix} s_1(\tau)e^{\lambda_{10}\tau} & s_2(\tau)e^{\lambda_{20}\tau} & \ldots & s_n(\tau)e^{\lambda_{n0}\tau} \end{pmatrix} T_{10}^{-1}(z(0, 0) - Z_0(0)).$$

We note that (H3a) is independent of the basis (37) chosen for the space of decaying solutions. Higher order terms in the boundary layer expansion (27) will also follow uniquely in turn since they will satisfy a nonhomogeneous version of the problem (34) with successively known, exponentially decaying forcing terms.

An alternative reformulation of the initial value problem (35) could be obtained by setting

(39) $$\rho_{20} = Km_{20}$$

where K is a constant symmetric solution of the algebraic Riccati problem

(40) $$KA_{40}(0) + A'_{40}(0)K - KB_{20}(0)B'_{20}(0)K + Q_{30}(0) = 0.$$

This is, of course, natural once we recognize (35) as an infinite interval nth order regulator problem which we shall call the initial boundary layer regulator. Assuming appropriate stabilizability-detectability assumptions (see, for example, Kučera [50] or [51]) would provide a unique positive semi-definite matrix K for which the remaining initial value problem for

(41) $$\frac{dm_{20}}{d\tau} = (A_{40}(0) - B_{20}(0)B'_{20}(0)K)m_{20}, \quad \tau \geq 0,$$

has a decaying solution. These hypotheses would, of course, be equivalent to (H3a) and somewhat weaker than the boundary layer controllability and observability assumptions of Wilde and Kokotović [87]. Because the origin is a saddle point for (35), direct numerical integration of (35) should be avoided. (One could not help but excite exponentially growing modes.) Numerical solution of (40)-(41) would be highly preferable. An alternative would be to follow Coppel [19]'s use of diagonalization of (35) via a nonsingular, symplectic matrix.

Proceeding analogously, we find that the terminal boundary layer correction (28) must satisfy the system (10) as a function of σ. Moreover, the boundary condition for q implies that it must also satisfy

(42) $\gamma_2(0, \varepsilon) - \Pi_3(\varepsilon)n(0, \varepsilon) \sim \Pi_2'(\varepsilon)\{X(1, \varepsilon)+\varepsilon n_1(0, \varepsilon)\} + \Pi_3(\varepsilon)Z(1, \varepsilon) - P_2(1, \varepsilon)$.

Continuing as before, we find that the leading terms will be a decaying solution of the system

(43) $$\frac{d}{d\sigma}\begin{pmatrix} n_{20} \\ \gamma_{20} \end{pmatrix} = -G(1)\begin{pmatrix} n_{20} \\ \gamma_{20} \end{pmatrix}$$

and

(44) $$\begin{cases} \frac{dn_{10}}{d\sigma} = -A_{20}(1)n_{20} + B_{10}(1)B_{20}'(1)\gamma_{20}, \\ \frac{d\gamma_{10}}{d\sigma} = Q_{20}(1)n_{20} + A_{30}'(1)\gamma_{20}, \end{cases}$$

while $\gamma_{20}(0) - \Pi_3(0)n_{20}(0)$ is determined by the limiting outer solution. Again (43) can be solved as a terminal boundary layer regulator and (44) would be integrated directly. We could also relate γ_{20} and n_{20} through the solution $\tilde{K}$ of an algebraic Riccati equation (see Sannuti [79]). Instead, let us take

(45) $$r_i(\sigma)e^{-\lambda_{i1}\sigma}, \quad i = 1, 2, \ldots, n,$$

to be n linearly independent decaying solutions to (43) as $\sigma \to \infty$ corresponding to eigenvalues λ_{i1} of $G(1)$ with positive real parts. Then we shall assume

(H3b) The $n \times n$ matrix $R_{21} - \Pi_3(0)R_{11}$ is nonsingular where

$$\begin{pmatrix} R_{11} \\ R_{21} \end{pmatrix} = \begin{pmatrix} r_1(0) & r_2(0) & \cdots & r_n(0) \end{pmatrix},$$

and it follows that the decaying solution of the initial value problem for (43) is uniquely given by

(46) $\begin{bmatrix} n_{20}(\sigma) \\ \gamma_{20}(\sigma) \end{bmatrix} = \begin{bmatrix} r_1(\sigma)e^{-\lambda_{11}\sigma} & r_2(\sigma)e^{-\lambda_{21}\sigma} & \cdots & r_n(\sigma)e^{-\lambda_{n1}\sigma} \end{bmatrix}$

$$\cdot (R_{21} - \Pi_3(0)R_{11})^{-1} (\Pi_2'(0)X_0(1) + \Pi_3(0)Z_0(1) - P_{20}(1)) .$$

Further terms also follow without difficulty.

D. FURTHER OBSERVATIONS

Under hypotheses (H1)-(H3), we have been able to formally obtain the terms of the asymptotic expansions (25) for the solution of the two-point problem (10) for the states and costates. The procedure is completely justified by the asymptotic theory for such problems (see, for example, Harris [34] or Vasil'eva and Butuzov [84]). To provide an independent proof, one would need to show that the difference between the solution and the $N + 1$ term approximation formally generated is $o(\epsilon^N)$ uniformly throughout $0 \le t \le 1$.

If the hypotheses used are not satisfied, the asymptotic solution to (10) will generally not have the form (25) and the limiting solution within (0, 1) may not satisfy the reduced problem (16)-(17).

As a boundary value problem for a linear system of dimension $2m + 2n$, (10) will have a unique solution for each fixed $\epsilon > 0$ if a determinant of that size is nonzero. Under our hypotheses, we have been able to essentially construct a fundamental matrix valid as $\epsilon \to 0$ and expand that determinant as an asymptotic power series in ϵ (see O'Malley and Keller [72]). The leading term of that determinant becomes the nonzero product of a $2m$th order determinant (nonzero by (H2)) and two nth order determinants (nonzero by (H3)). Assumption (H1) guarantees that any limiting solution within (0, 1) will satisfy the reduced system (16), although without (H2) and (H3), it could blow up like, for example, $\epsilon^{-\kappa}$, $\kappa > 0$. As long as G remained nonsingular, we could define the reduced system (16), but if $G(t)$ had purely imaginary eigenvalues, the solution might be rapidly oscillating (like $\sin t/\epsilon$) as $\epsilon \to 0$. A different type of asymptotic analysis, combined with boundary layer theory, would then be required (see, for example, Hoppensteadt and Miranker [37]).

Under our hypotheses, it is relatively easy to write explicit expressions for the limiting solutions

(47) $\begin{cases} x(t, \epsilon) = X_0(t) + O(\epsilon) , \\ z(t, \epsilon) = Z_0(t) + m_{20}(\tau) + n_{20}(\sigma) + O(\epsilon) , \\ u(t, \epsilon) = U_0(t) + v_0(\tau) + w_0(\sigma) + O(\epsilon) , \text{ and} \\ J^*(\epsilon) = J_0 + O(\epsilon) . \end{cases}$

Indeed, it is convenient to refer to X_0 as the "slow state" x_s, with the corresponding Riccati matrix K_{10} (see (20)) called the slow Riccati gain K_s. Together these determine the slow variables $z_s = Z_0$, $u_s = U_0$, and $J_s = J_0$. Likewise, we can call m_{20} the fast initial state z_{f0} and K (see (40)) the fast initial (boundary layer) Riccati gain K_{f0}. Then we will have the corresponding fast initial control $u_{f0}(\tau) = v_0(\tau) = -B_2'(0) K_{f0} z_{f0}(\tau)$. Doing the same for the fast terminal transients, we would rewrite (47) as

(48)
$$\begin{cases} x(t, \varepsilon) = x_s(t) + O(\varepsilon), \\ z(t, \varepsilon) = z_s(t) + z_{f0}(\tau) + z_{f1}(\sigma) + O(\varepsilon), \\ u(t, \varepsilon) = u_s(t) + u_{f0}(\tau) + u_{f1}(\sigma) + O(\varepsilon), \text{ and} \\ J^*(\varepsilon) = J_s + O(\varepsilon). \end{cases}$$

Since dz/dt and du/dt are $O(1/\varepsilon)$ near the endpoints, while $dx/dt = O(1)$ everywhere, our earlier reference to z and u as fast-variables and to x as a slow-variable is justified.

This three-time scale separation (see Chow and Kokotović [13] and Chow [12]) is valuable for design purposes. It reflects the intuitively desirable idea of a three stage design process consisting of a slow system (that is, the mth order regulator problem (19) for $0 \leq t \leq 1$) improved by two separate fast systems (that is, the nth order boundary layer regulator problems (35) and (43) which are infinite interval problems in the stretched variables $\tau = t/\varepsilon$ and $\sigma = (1-t)/\varepsilon$, respectively). The fast systems correct the lower dimensional slow system at the endpoints $t = 0$ and $t = 1$. The time-scale separation becomes more apparent after a preliminary transformation of the system (10) to diagonal form (see Chang [11]), but some interaction between time scales (as in our construction of the formal solution to (25)) is needed to analyze higher order approximate solutions.

We note that some care must be exercised in applying these results. Wilde and Kokotović [87], for example, observe that if A_{40} is not stable, difficulties could result if an asymptotic approximation to the optimal control is inserted into the state equations (2) and the result is integrated. This relates to the usual problem of sensitivity regarding open loop control, but the difficulty can be avoided by only using the asymptotic formulas already developed or by combining open and closed loop control as Wilde and Kokotović suggest.

Many practical problems concerning the use of singular perturbation theory for such regulator problems remain unanswered. Since our results are asymptotic, how small should ε be in order to use these results? If some time constants are much smaller than others, should we instead use a more refined model like the system

$$\begin{cases} \dfrac{dx}{dt} = A_1(t, \varepsilon, \mu)x + A_2(t, \varepsilon, \mu)z + A_3(t, \varepsilon, \mu)w + B_1(t, \varepsilon, \mu)u \ , \\[6pt] \varepsilon \dfrac{dz}{dt} = A_4(t, \varepsilon, \mu)x + A_5(t, \varepsilon, \mu)z + A_6(t, \varepsilon, \mu)w + B_2(t, \varepsilon, \mu)u \ , \\[6pt] \varepsilon\mu \dfrac{dw}{dt} = A_7(t, \varepsilon, \mu)x + A_8(t, \varepsilon, \mu)z + A_9(t, \varepsilon, \mu)w + B_3(t, \varepsilon, \mu)u \ ? \end{cases}$$

Neglecting to raise further important questions, we are nonetheless relatively content with our conclusions which we now summarize.

THEOREM. *For the problem (1)-(4), suppose*

(i) *all eigenvalues of the Hamiltonian matrix*

$$G(t) = \begin{pmatrix} A_{40} & -B_{20}B'_{20} \\ -Q_{30} & -A'_{40} \end{pmatrix}$$

have nonzero real parts throughout $0 \le t \le 1$,

(ii) *the reduced (or outer) mth order regulator problem*

$$\begin{cases} \dfrac{d}{dt}\begin{pmatrix} X_0 \\ P_{10} \end{pmatrix} = \begin{pmatrix} V_1 & V_2 \\ V_3 & -V'_1 \end{pmatrix} \begin{pmatrix} X_0 \\ P_{10} \end{pmatrix} , \\[10pt] X_0(0) = x(0, 0) \ , \quad P_{10}(1) = \Pi_1(0)X_0(1) \end{cases}$$

has a unique solution,

(iiia) *the initial nth order boundary layer regulator*

$$\dfrac{d}{d\tau}\begin{pmatrix} m_{20} \\ \rho_{20} \end{pmatrix} = G(0)\begin{pmatrix} m_{20} \\ \rho_{20} \end{pmatrix}$$

has a unique decaying solution on $0 \le \tau < \infty$ *for any initial value* $m_{20}(0)$,

(iiib) *the terminal nth order boundary layer regulator*

$$\dfrac{d}{d\sigma}\begin{pmatrix} n_{20} \\ \gamma_{20} \end{pmatrix} = -G(1)\begin{pmatrix} n_{20} \\ \gamma_{20} \end{pmatrix}$$

has a unique decaying solution on $0 \le \sigma < \infty$ *for any value* $\gamma_{20}(0) - \Pi_3(0)n_{20}(0)$.

Then the problem has a unique solution with the optimal control, the corresponding trajectories, and the optimal cost being of the form

$$\begin{cases} u(t, \varepsilon) = U(t, \varepsilon) + v(\tau, \varepsilon) + w(\sigma, \varepsilon) , \\ x(t, \varepsilon) = X(t, \varepsilon) + \varepsilon m_1(\tau, \varepsilon) + \varepsilon n_1(\sigma, \varepsilon) , \\ z(t, \varepsilon) = Z(t, \varepsilon) + m_2(\tau, \varepsilon) + n_2(\sigma, \varepsilon) , \\ J^*(\varepsilon) \sim \sum_{k=0}^{\infty} J_k \varepsilon^k , \end{cases}$$

in $0 \le t \le 1$ where the terms all have asymptotic power series expansions and the functions of $\tau = t/\varepsilon$ or $\sigma = (1-t)/\varepsilon$ decay to zero as that variable tends to infinity.

4. The Nonlinear Singularly Perturbed Regulator Problem

A. INTRODUCTION

Several investigators, including McIntire [58] whom we shall largely follow, have considered the nonlinear problem of minimizing the scalar cost

(1) $$J = \lambda\big(x(1), z(1), \varepsilon\big) + \int_0^1 \Lambda\big(x(t), z(t), u(t), t, \varepsilon\big) dt$$

subject to the state constraints

(2) $$\begin{cases} \dot{x} = f(x, z, u, t, \varepsilon) , \quad x(0) \text{ given}, \\ \varepsilon \dot{z} = g(x, z, u, t, \varepsilon) , \quad z(0) \text{ given}, \end{cases}$$

where x, z, and u are vectors of dimensions m, n, and r, respectively, and ε is a small positive parameter.

Necessary conditions for optimality follow under mild assumptions, for example, from the Pontryagin Maximum Principle and corresponding variational arguments. If we define the Hamiltonian

(3) $$h = \Lambda + p'f + q'g$$

where the costates p and εq satisfy

(4) $$\begin{cases} \dot{p} = -h_x , \quad p(1, \varepsilon) = \lambda_x\big(x(1), z(1), \varepsilon\big) , \\ \varepsilon \dot{q} = -h_z , \quad \varepsilon q(1, \varepsilon) = \lambda_z\big(x(1), z(1), \varepsilon\big) , \end{cases}$$

and the optimality condition becomes

(5) $$h_u = 0$$

while minimization will require the Legendre-Clebsch condition

$$h_{uu} \geq 0,$$

that is, this hessian matrix must be positive semi-definite at least locally.

To provide a bounded terminal value for q, we will ask that λ be a slowly-varying function of the fast state variable z, that is,

(6) $$\lambda(x, z, \varepsilon) = \theta(x, \varepsilon z, \varepsilon).$$

Then the terminal cost λ will depend only on the slow state x at $\varepsilon = 0$, so that both z and u will play the role of control vectors in the reduced control problem obtained by setting $\varepsilon = 0$ in (1) and (2).

Our results for the linear problem suggest that it is more convenient to introduce the Hamiltonian

(7) $$H(\psi, \zeta, u, t, \varepsilon) = h(x, z, u, p, q, t, \varepsilon)$$

for the vectors

$$\psi = \binom{x}{p} \quad \text{and} \quad \zeta = \binom{z}{q}$$

of dimensions $2m$ and $2n$ respectively. The necessary conditions for optimality then reduce to a two-point boundary value problem for the nonlinear singularly perturbed system

(8) $$\begin{cases} \dot{\psi} = J_m H_\psi(\psi, \zeta, u, t, \varepsilon) \\ \varepsilon \dot{\zeta} = J_n H_\zeta(\psi, \zeta, u, t, \varepsilon) \\ 0 = H_u(\psi, \zeta, u, t, \varepsilon), \end{cases}$$

where $J_k = \begin{pmatrix} 0 & I_k \\ -I_k & 0 \end{pmatrix}$, $H_{uu} \geq 0$, and $x(0)$ and $z(0)$ are prescribed while

$$p(1, \varepsilon) = \theta_x(x(1, \varepsilon), \varepsilon z(1, \varepsilon), \varepsilon) \quad \text{and} \quad q(1, \varepsilon) = \frac{\partial \theta}{\partial (\varepsilon z)}(x(1, \varepsilon), \varepsilon z(1, \varepsilon), \varepsilon).$$

Henceforth, then, we shall restrict attention to the asymptotic solution of the boundary value problem for (8).

B. THE STRONG LEGENDRE-CLEBSCH CONDITION

Let us assume the strong form of the Legendre-Clebsch condition, that is,

(9) $$H_{uu} > 0.$$

(We shall not be cautious in defining a limit to the region where the hessian is positive-definite, nor to limiting the smoothness of functions in (8), though we realize that such restrictions would be of practical importance.) Under (9), the implicit function theorem implies that we can uniquely solve the optimality condition $H_u = 0$ for

(10) $$u = \eta(\psi, \zeta, t, \varepsilon).$$

Thus (8) reduces to the singularly perturbed boundary value problem

(11)
$$\begin{cases} \dot{\psi} = F(\psi, \zeta, t, \varepsilon) \equiv J_m H_\psi(\psi, \zeta, \eta, t, \varepsilon) \\ \varepsilon \dot{\zeta} = G(\psi, \zeta, t, \varepsilon) \equiv J_n H_\zeta(\psi, \zeta, \eta, t, \varepsilon), \end{cases}$$

together with the $2m + 2n$ separated boundary conditions inherited from (2), (4) and (6). (If $H_{uu} > 0$ only held locally several roots $u = \eta$ might be possible, leading to different two-point problems (11).)

Regretably, no adequate theory is available for such singularly perturbed two-point problems. For initial value problems, Tikhonov developed a theory under the assumption that all eigenvalues of G_ζ are locally stable (see Vasil'eva and Butuzov [84]). One would expect to be able to solve appropriate boundary value problems when G_ζ has a fixed number of stable eigenvalues and a fixed number of unstable ones. We find, however, a need to restrict nonlinearities. One might, for example, expect the solution of the two-point problem

$$\begin{cases} \dot{u} = v, \quad u(0) = 0, \quad u(1) = 1, \\ \varepsilon \dot{v} = -v - v^3, \end{cases}$$

to converge to the limit $(U_0, V_0) = (1, 0)$ for $t > 0$, just as the limiting solution of the initial value problem

$$\begin{cases} \dot{u} = v, \quad u(0) = 0, \\ \varepsilon \dot{v} = -v - v^3, \quad v(0) = 1, \end{cases}$$

is $(U_0, V_0) = (0, 0)$. One can easily show, however, (see Coddington and Levinson [16]) that the first problem has no solution for ε small. Such examples have effectively limited most singular perturbation analysis to problems like (11) with G linear or quadratic in the fast-variable ζ.

If a limiting solution to our control problem exists, we can nonetheless expect it to satisfy the reduced problem

(12)
$$\begin{cases} \dot{\Psi}_0 = F(\Psi_0, Z_0, t, 0), \\ 0 = G(\Psi_0, Z_0, t, 0), \\ X_0(0) = x(0, 0), \quad P_0(1) = \theta_x(X_0(1), 0, 0), \end{cases}$$

where $\Psi_0 = \begin{pmatrix} X_0 \\ P_0 \end{pmatrix}$ and $Z_0 = \begin{pmatrix} \tilde{Z}_0 \\ Q_0 \end{pmatrix}$. Since the initial condition for z and the terminal condition for q have been neglected, we will generally need boundary layers (that is, nonuniform convergence of the solution) near both the endpoints $t = 0$ and $t = 1$.

Corresponding to any root

(13)
$$Z_0 = \xi(\Psi_0, t)$$

of $G(\Psi_0, \xi, t, 0) = 0$, we obtain a reduced regulator problem

(14)
$$\begin{cases} \dot{\Psi}_0 = F(\Psi_0, t) \equiv F(\Psi_0, \xi(\Psi_0, t), t, 0), \\ X_0(0) = x(0, 0), \quad P_0(1) = \theta_x(X_0(1), 0, 0). \end{cases}$$

It would be natural to assume

(H-a) *the reduced regulator problem (14) has a unique solution for* $0 \leq t \leq 1$.
(We recall that the corresponding hypothesis (H2) for the linear problem was far more concrete and easily verified.)

We note that the important matrix G_ζ is given by $G_\zeta = J_n\left[H_{\zeta\zeta} + H_{\zeta u}\frac{\partial\eta}{\partial\zeta}\right]$. Moreover, differentiation of $H_u \equiv 0$ implies that $H_{uu}\frac{\partial\eta}{\partial\zeta} + H_{u\zeta} = 0$, so $H_{\zeta u} = H'_{u\zeta}$ implies that

(15)
$$G_\zeta = J_n\left[H_{\zeta\zeta} - H'_{u\zeta}H_{uu}^{-1}H_{u\zeta}\right]$$

and the symmetry of $J_n G_\zeta$ implies that the eigenvalues of G_ζ occur in pairs $\pm \lambda$. We can therefore guarantee that the $2n \times 2n$ matrix G has n stable eigenvalues and n unstable ones if we assume that for all arguments ψ, ζ,

(H-b) G_ζ *has no purely imaginary eigenvalues on* $0 \leq t \leq 1$ *for* $\varepsilon = 0$.

This strong assumption implies that G_ζ is nonsingular, so there would be only one root Z_0 of $G = 0$. If it held only locally, more roots would be possible.

We might now proceed as for the linear problem and seek an asymptotic solution to our two-point boundary value problem (11) in the form

(16)
$$\begin{cases} \psi(t, \varepsilon) = \Psi(t, \varepsilon) + \varepsilon\psi_L(\tau, \varepsilon) + \varepsilon\psi_R(\sigma, \dot{\varepsilon}), \\ \zeta(t, \varepsilon) = Z(t, \varepsilon) + \zeta_L(\tau, \varepsilon) + \zeta_R(\sigma, \varepsilon), \end{cases}$$

where all terms have power series expansions in ε and the functions of the stretched variables

$$\tau = t/\varepsilon \quad \text{or} \quad \sigma = (1-t)/\varepsilon$$

decay to zero as that stretched variable tends to infinity. Corresponding expansions for the optimal control and optimum cost will follow through (10) and (1), respectively. Such expansions can be shown (under hypotheses (H-a), (H-b), and (H-c) (below)) to provide the unique asymptotic solution when f and g are linear in z and u and Λ is quadratic in these variables (see, for example, O'Malley [64]). They may be valid more generally, though the appropriate stretched variables may be

$$\tau' = t/e^{-1/\varepsilon} \quad \text{and} \quad \sigma' = (1-t)/e^{-1/\varepsilon}$$

when the two-point problem (8) is quadratic in ζ (see Višik and Lyusternik [85]).

When (16) holds, the solution will be asymptotically provided by the outer expansion

(17) $$\left(\Psi(t, \varepsilon), Z(t, \varepsilon)\right) \sim \sum_{j=0}^{\infty} \left(\Psi_j, Z_j\right)\varepsilon^j$$

within $(0, 1)$. It will therefore satisfy the full system (11) as a power series in ε. The leading term $\left(\Psi_0, Z_0\right)$ will satisfy the reduced problem (12) and, under hypotheses (H-a), (H-b), be the unique solution of (13)-(14). The next term will satisfy the linear problem obtained by linearization of (14) about $\left(\Psi_0, Z_0, t, 0\right)$, namely

$$\begin{cases} \dot{\Psi}_1 = F_{0\psi}\Psi_1 + F_{0\zeta}Z_1 + F_{0\varepsilon}, \\ \dot{Z}_0 = G_{0\psi}\Psi_1 + G_{0\zeta}Z_1 + G_{0\varepsilon}, \quad \text{with} \\ X_1(0) \text{ and } P_1(1) - \theta_{xx}(X_0(1), 0, 0)X_1(1) \text{ known successively,} \end{cases}$$

where, for example, $F_{0\psi} = \frac{\partial F}{\partial \psi}\left(\Psi_0, Z_0, t, 0\right)$. Since $G_{0\zeta}$ is invertible, we can solve the second equation for Z_1 leaving a linear regulator problem

(18) $$\dot{\Psi}_1 = W_0\Psi_1 + w_0$$

for Ψ_1 with

$$W_0 \equiv F_{0\psi} - F_{0\zeta}G_{0\zeta}^{-1}G_{0\psi} = \frac{\partial F}{\partial \psi}\left(\Psi_0, t\right).$$

Using the definitions of F and G plus (15), further manipulation implies that

$$W_0 = J_m\left\{\left(H_{\psi\psi} - H'_{u\psi}H_{uu}^{-1}H_{u\psi}\right) + \left(H_{\psi\zeta} - H'_{u\psi}H_{uu}^{-1}H_{u\zeta}\right)'\left(H_{\zeta\zeta} - H'_{u\zeta}H_{uu}^{-1}H_{u\zeta}\right)^{-1}\left(H_{\psi\zeta} - H'_{u\psi}H_{uu}^{-1}H_{u\zeta}\right)\right\}$$

and the symmetry of $J_m W_0$ implies that W_0 has the form

(19) $$W_0 = \begin{pmatrix} W_{11} & W_{12} \\ W_{21} & -W_{11}' \end{pmatrix}$$

for symmetric matrices W_{12} and W_{21}. By hypothesis (H-a), the linearized homogeneous problem corresponding to (14) has a unique solution. Thus the Fredholm alternative implies that the same is true for (18). On the other hand, at least for $\theta_{xx}(X_0(1), 0, 0)$ positive semidefinite, Bucy [6] suggests that uniqueness of (18) implies that W_{12} and W_{21} are negative semidefinite and (18) could be solved through a matrix Riccati feedback.

Near $t = 0$, σ is asymptotically infinite, so we have $\psi \sim \Psi + \varepsilon \psi_L$, $\frac{d\psi}{dt} \sim \frac{d\Psi}{dt} + \frac{d\psi_L}{d\tau}$, and so on. The nonlinear system (11) then implies that the initial boundary layer correction $\begin{pmatrix} \varepsilon \psi_L \\ \zeta_L \end{pmatrix}$ must be a decaying solution of

(20) $$\begin{cases} \dfrac{d\psi_L}{d\tau} = F(\Psi+\varepsilon\psi_L, Z+\zeta_L, \varepsilon\tau, \varepsilon) - F(\Psi, Z, \varepsilon\tau, \varepsilon), \\ \dfrac{d\zeta_L}{d\tau} = G(\Psi+\varepsilon\psi_L, Z+\zeta_L, \varepsilon\tau, \varepsilon) - G(\Psi, Z, \varepsilon\tau, \varepsilon), \end{cases}$$

with

(21) $$z_L(0, \varepsilon) \sim z(0, \varepsilon) - \tilde{Z}(0, \varepsilon)$$

for $\zeta_L = \begin{pmatrix} z_L \\ q_L \end{pmatrix}$ split in half. Thus this boundary layer will account for the boundary value for z. When $\varepsilon = 0$, then we seek a decaying solution of the system

(22) $$\begin{cases} \dfrac{d\psi_{L0}}{d\tau} = F(\Psi_0(0), Z_0(0)+\zeta_{L0}, 0, 0) - F(\Psi_0(0), Z_0(0), 0, 0), \\ \dfrac{d\zeta_{L0}}{d\tau} = G(\Psi_0(0), Z_0(0)+\zeta_{L0}, 0, 0), \text{ with} \\ z_{L0}(0) = z(0, 0) - \tilde{Z}_0(0), \end{cases}$$

since $G(\Psi_0(0), Z_0(0), 0, 0) = 0$. Presuming we can find an exponentially decaying solution ζ_{L0} of the second system, it will determine $d\psi_{L0}/d\tau$ and thereby

$$\psi_{L0}(\tau) = -\int_\tau^\infty \frac{d\psi_{L0}(s)}{d\tau} ds.$$

We are now left with a nonlinear infinite interval boundary layer regulator problem for ζ_{L0}, namely,

(23)
$$\begin{cases} \dfrac{d\zeta_{L0}}{d\tau} = G\bigl(\Psi_0(0),\, Z_0(0)+\zeta_{L0},\, 0,\, 0\bigr) - G\bigl(\Psi_0(0),\, Z_0(0),\, 0,\, 0\bigr)\,, & \tau \geq 0\,, \\ z_{L0}(0) \text{ prescribed.} \end{cases}$$

Here $\zeta_{L0} = 0$ is a rest point of the system and G_ζ has (under hypothesis (H-b)) n stable eigenvalues and n unstable ones. The origin is a saddle point for the system and we face a classical problem of conditional stability. Standard theory (see, for example, Coddington and Levinson [17]) implies that there is an n dimensional manifold of initial values $\zeta_{L0}(0)$ (near the origin) resulting in a decaying solution to (23). We need hypotheses which guarantee that the prescribed n vector $z_{L0}(0)$ lies on this manifold for some choice of the last components $q_{L0}(0)$ of $\zeta_{L0}(0)$ (see Levin [55] who first encountered such problems in singular perturbations). This is guaranteed by our hypothesis (H3-a) for the linear problem and such an assumption may also be reasonable for nonlinear problems which have (very) small "boundary layer jumps" $|z(0,0)-\tilde{Z}_0(0)|$ (see Hadlock [33]). To avoid this restrictive assumption for corresponding boundary value problems, Vasil'eva and Butuzov [84] require the stable manifold be describable in the form

$$q_{L0} = \phi^L(z_{L0})$$

for some function ϕ^L. This reduces the problem (23) to an nth order initial value problem for z_{L0} for which the origin is asymptotically stable. Other possible approaches include Liapunov functions (see Habets [31]). Analogous, but linear, problems occur in obtaining higher order terms $\begin{pmatrix}\Psi_{Lj}\\\zeta_{Lj}\end{pmatrix}$, $j > 0$, and the problems recur in obtaining the terminal boundary layer correction $\begin{pmatrix}\varepsilon\Psi_R(\sigma,\varepsilon)\\\zeta_R(\sigma,\varepsilon)\end{pmatrix}$.

Thus we ask

(H-c) *the initial boundary layer regulator*

$$\frac{d\zeta_{L0}}{d\tau} = G\bigl(\Psi_0(0),\, Z_0(0)+\zeta_{L0},\, 0,\, 0\bigr)\,, \quad z_{L0}(0) \text{ prescribed,}$$

has a unique decaying solution for $\tau \geq 0$, *and the terminal boundary layer regulator*

$$\frac{d\zeta_{R0}}{d\sigma} = -G\bigl(\Psi_0(1),\, Z_0(1)+\zeta_{R0},\, 1,\, 0\bigr)\,, \quad q_{R0}(0) = \frac{\partial\theta}{\partial(\varepsilon z)}\bigl(X_0(1),\, 0,\, 0\bigr) - Q_0(1)\,,$$

has a unique decaying solution for $\sigma \geq 0$.

As mentioned earlier, these are the correct assumptions when specialized to the quasilinear problem. Much further work needs to be done, even for the restricted problem when the strong Legendre-Clebsch condition holds.

C. THE WEAK LEGENDRE-CLEBSCH CONDITION

A more difficult, but more interesting, problem occurs if we let H_{uu} become singular. Then we might have multiple solutions $u = \eta_i(\psi, \zeta, t, \varepsilon)$ of the optimality condition $H_u = 0$ and we can anticipate the possibility of switching back and forth between them. (This would seem especially likely to occur if $\det(H_{uu}) = 0$ at isolated points where one root loses stability to another.) One would expect the states and costates to have corners at these interior transition points of the control, similar to behavior with bang-bang control (see Freedman and Kaplan [26] and Kokotović and Haddad [47]). More impulsive control would lead to jumps in the states and costates. If $H_{uu} \equiv 0$ along a trajectory, a singular arc occurs and further necessary conditions for optimality can be obtained by differentiation (see Bell and Jacobson [4]).

Simple singular perturbation problems featuring such discontinuous solutions include

$$\begin{cases} \dot{u} = v, & u(0) = 0, \quad u(1) = \tfrac{1}{8}, \\ \varepsilon \dot{v} = v - v^3. \end{cases}$$

This problem (almost like the Coddington and Levinson nonexistence example cited above) has the limiting solution

$$u(t) = \begin{cases} 0, & 0 \leq t \leq \tfrac{1}{2}, \\ t - \tfrac{1}{2}, & \tfrac{1}{2} \leq t \leq 1, \end{cases}$$

so that v jumps discontinuously at $t = \tfrac{1}{2}$ as $\varepsilon \to 0$. We note that the reduced problem is satisfied everywhere, but the root changes at $t = \tfrac{1}{2}$. Pictorially, we have

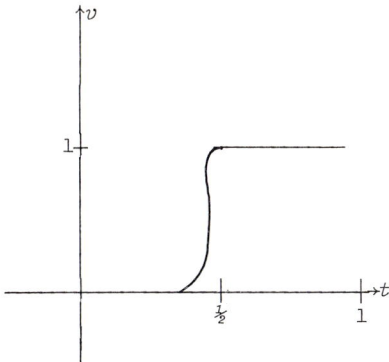

The example

$$\begin{cases} \dot{u} = v, & -1 \le t \le 1, \quad u(-1) = -2, \quad u(1) = -1, \\ \varepsilon \dot{v} = -tv^2 + u, \end{cases}$$

can be shown (see Howes [39]) to have a unique limiting solution

$$u(t) = \begin{cases} -(\sqrt{2}-1+\sqrt{-t})^2, & -1 \le t < 0, \\ 0, & 0 < t < 1, \end{cases}$$

with $v = \dot{u}$. We note that the reduced system $tv^2 = u$ is satisfied for $t \ne 0, 1$ and that the behavior at $t = 0$ is reminiscent of jump phenomena which occurs at gas dynamical shocks and for bang-bang control. There is also an ordinary boundary layer at $t = 1$. For small ε, the solution is as in the figure.

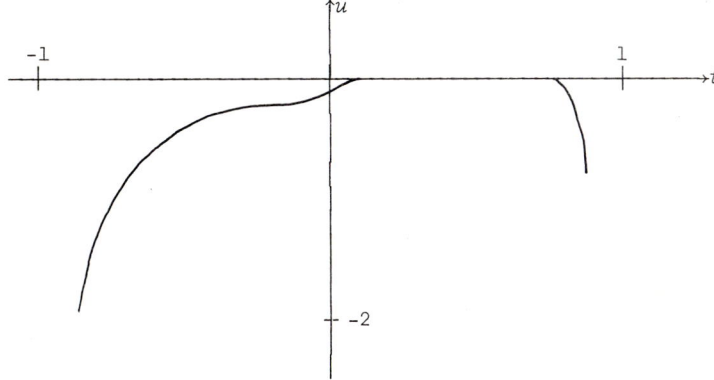

Studying several roots of $H_u = 0$ simultaneously recalls the geometric theories of Levinson [56] and Andronov, Vitt, and Khaikin [2] for higher order generalizations of the van der Pol oscillator, Fife's recent work on "transition layers" occurring as stationary patterns for reaction-diffusion systems (see Fife [23]), and Carpenter's work using isolating blocks for autonomous systems to provide existence theorems and asymptotic limits to the Fitzhugh-Nagumo and other nerve impulse equations (see Carpenter [9]).

The need for a global analysis of such problems is clear, both to study the stable solution manifolds involved and their regions of attraction. It would seem profitable to specialize the study to singularly perturbed Hamiltonian systems, rather than more general boundary value problems. The most complete discussion for nonlinear problems is contained in Vasil'eva and Butuzov [84], but that is inadequate for our problem. One new approach, differential inequalities, has been successfully used by Howes [38], [40], to obtain existence and asymptotic behavior for scalar problems $\varepsilon \ddot{y} = F(t, y, \dot{y}, \varepsilon)$ with F quadratic in $\dot{y}$. Likewise, Chueh, Conley, and Smoller [15] have used invariant rectangles to obtain comparison theorems for higher order systems. It remains much easier to ask relevant questions than to answer them.

5. Cheap Control Theory and Singular Arcs

A. THE GENERAL PROBLEM

Now consider a linear regulator problem with cheap control, that is, let us seek to minimize

$$(1) \qquad J(\varepsilon) = \tfrac{1}{2} \int_0^1 \left[x'Q(t)x + \varepsilon^2 u'R(t)u \right] dt$$

where ε is a small positive parameter subject to the (initial value problem) constraint

$$(2) \qquad \dot{x} = A(t)x + B(t)u, \quad 0 \leq t \leq 1, \quad x(0) \text{ prescribed.}$$

We shall take Q and R to be symmetric positive semidefinite and positive definite matrices, respectively, and we shall assume that Q, R, A, and B are infinitely differentiable functions of t. Our presentation will rely heavily on joint work with Antony Jameson.

Such problems arise naturally in a variety of contexts. Kwakernaak and Sivan [53], Wonham [90], Kwakernaak [52], and Francis and Glover [25] were concerned with the asymptotic location of poles for closed loop systems; Friedland [27] and Moylan [59] examined the limiting possibilities for filters; Kalman [45] and Moylan and Anderson [60] utilized related problems to study inverse optimal control problems; and Lions [57] studied analogous cheap control problems (that is, those for which the control is cheap relative to the state in the cost functional) for distributed parameter systems.

When $\varepsilon = 0$, (1)-(2) becomes a singular problem of optimal control. Our asymptotic analysis of (1)-(2) allows study of the singular problem as the (nonuniform) limit of the nearly singular problems with small $\varepsilon > 0$. Indeed, it provides an important new tool for analyzing singular optimal control problems (see

Bell and Jacobson [4]). This idea was used earlier in the singular control literature by Jacobson and coworkers to both theoretical and numerical advantage (see Jacobson and Speyer [42], Jacobson, Gershwin, and Lele [41], and Coppel [20]) and is analogous to the common usage of artificial viscosity techniques in fluid dynamics (see, for example, Richtmyer and Morton [76]).

For each $\varepsilon > 0$, standard theory (see, for example, Anderson and Moore [1]) implies that the optimal control is given by

(3) $$u = -\frac{1}{\varepsilon^2} R^{-1} B' p$$

where the costate vector p satisfies the system adjoint to that for x. (The equation (3) suggests a more general asymptotic study of high gain control systems via singular perturbation methods (see Young, Kokotovic, and Utkin [92] and O'Malley [67]).) Eliminating u from the state equation, then, results in the linear, singularly perturbed boundary value problem

(4) $$\begin{cases} \varepsilon^2 \dot{x} = \varepsilon^2 A x - B R^{-1} B' p, & x(0) \text{ prescribed,} \\ \\ \dot{p} = -Q x - A' p, & p(1) = 0. \end{cases}$$

Because $\varepsilon^2 A$ is singular when $\varepsilon = 0$, the familiar singular perturbation theory cited earlier (as in Harris [34]) does not apply. One might transform the problem to a (higher dimensional) problem where those theorems are applicable. This was done by O'Malley and Jameson [70], [71], using transformations like those found in the earlier control papers of Moylan and Moore [61] or Friedland [27]. We will present a more direct solution here.

To anticipate our general results, consider the simplest such scalar problem with

(5) $$\begin{cases} J(\varepsilon) = \tfrac{1}{2} \int_0^1 (x^2 + \varepsilon^2 u^2) dt, \\ \\ \dot{x} = u, \quad x(0) = 1. \end{cases}$$

The exact solution of the resulting two-point problem provides the optimal trajectory

$$x(t, \varepsilon) = \left(e^{-t/\varepsilon} + e^{-1/\varepsilon} e^{-(1-t)/\varepsilon} \right) / \left(1 + e^{-2/\varepsilon} \right)$$

and the optimal control $u = \dot{x}$. Since $e^{-1/\varepsilon}$ is asymptotically negligible as $\varepsilon \to 0$, we have

$$\bigl(x(t, \varepsilon), u(t, \varepsilon)\bigr) \approx \left(e^{-t/\varepsilon}, -\frac{1}{\varepsilon} e^{-t/\varepsilon} \right).$$

Both functions tend to zero as $\varepsilon \to 0$ for $t > 0$, but they converge nonuniformly at

$t = 0$. Indeed, $\frac{1}{\varepsilon} e^{-t/\varepsilon}$ behaves like a delta function peaked at $t = 0$ in the limit $\varepsilon \to 0$, while $e^{-t/\varepsilon}$ behaves like $1 - H$ for the Heaviside function H on $t \geq 0$ as $\varepsilon \to 0$. The initial impulse drives the state to zero instantaneously, in agreement with the intuitive answer $u = -\delta$ long known (see Ho [36]). We note that the optimal cost satisfies

$$J^*(\varepsilon) \approx \int_0^1 e^{-2t/\varepsilon} dt = O(\varepsilon),$$

that is, it is asymptotically cost free.

More generally, we can expect the solution to our two-point problem (4) to feature endpoint regions of nonuniform convergence, with convergence elsewhere to a solution of the reduced equation

$$BR^{-1}B'P_0 = 0.$$

Indeed, it is well known that the solution to the corresponding singular control problem follows a (usually lower dimensional) singular arc (along which $B'P_0 = 0$). If B^{-1} exists, $P_0 = 0$ and, then, if Q^{-1} exists, $X_0 = 0$. Otherwise, as in bifurcation theory (see Cesari [10]), an auxiliary equation is needed to determine a unique limiting solution. Here $B'P_0 = 0$ restricts P_0 to a lower dimensional space. The resulting nonuniform convergence is, of course, the hallmark of singular perturbation problems. Our singular perturbation analysis, however, provides us with the unique limiting solution along the singular arc as well as the appropriate impulse at $t = 0$ to get us onto this arc.

The results we shall obtain differ fundamentally in a hierarchy of cases. Thus, we define Case L, $L \geq 1$, by requiring that

(6)
$$\begin{cases} B_j' Q B_j = 0, & j = 0, 1, \ldots, L-2, \\ B_{L-1}' Q B_{L-1} > 0, \end{cases}$$

throughout $0 \leq t \leq 1$ where

(7)
$$B_0 = B \quad \text{and} \quad B_j = AB_{j-1} - \dot{B}_{j-1}, \quad j \geq 1.$$

This corresponds to the usual definition of singular arcs of order L (see Goh [29] and Robbins [77]). There are, of course, problems between cases, ones where the case changes with t, and ones beyond all cases (such as $Q \equiv 0$). Nonetheless, in Case L, we find that the optimal control takes the form

(8) $$u(t, \varepsilon) = U(t, \mu) + \frac{1}{\mu^L} v(\tau, \mu) + w(\sigma, \mu)$$

while the corresponding trajectory is like

(9) $$x(t, \varepsilon) = X(t, \mu) + \frac{1}{\mu^{L-1}} m(\tau, \mu) + \mu n(\sigma, \mu)$$

where the series are in powers of

$$\mu = \varepsilon^{1/L}$$

and the stretched variables providing endpoint boundary layers are

$$\tau = t/\mu \quad \text{and} \quad \sigma = (1-t)/\mu.$$

The corrections v and m (and w and n) tend to zero as τ (and σ) tend to infinity. It is most important to note that the optimal control features an initial impulse

$$\frac{1}{\mu^L} \sum_{j=0}^{L-1} v_j(\tau) \mu^j$$

which we will find behaves like a linear combination of matrix impulse functions $\delta, \delta', \ldots, \delta^{(L-1)}$ with δ behaving like the asymptotic limit of a matrix $\frac{C}{\mu} e^{-Ct/\mu}$ as $\mu \to 0$ for a positive definite matrix C. Such an impulse will allow a rapid transfer from the given initial state $x(0)$ in n-space to $X(0^+, 0)$ on a singular arc lying on a manifold of dimension $n - Lr$. The limiting control $U(t, 0)$ within $(0, 1)$ is that corresponding to the singular arc solution. At $t = 1$, the control will converge nonuniformly, but it will not be impulse-like. We note that the trajectory will feature impulsive behavior at $t = 0$ whenever $L > 1$ and that the optimal cost $J^*(\mu)$ will have an asymptotic power series expansion in μ whose limit $J_0 = \frac{1}{2} \int_0^1 X'(t, 0) Q(t) X(t, 0) dt$ is the cost of following the singular arc solution.

For more details see O'Malley and Jameson [71].

B. CASE 1

We will now limit attention to Case 1 problems where

(10) $$B'QB > 0$$

in $[0, 1]$, noting that it is the only case with bounded state x at $t = 0$. (A different frequency domain condition for "bounded peaking" is given by Francis and Glover [25].) Since a straightforward presentation of the state-costate solution is given by O'Malley [66], we will instead seek a solution with $p = kx$ so that the optimal control will be in the feedback form

(11) $$u = -\frac{1}{\varepsilon^2} R^{-1} B' k x$$

where the matrix k is the unique symmetric, positive semi-definite solution of the $n \times n$ Riccati terminal value problem

(12) $$\varepsilon^2 \dot{k} + \varepsilon^2(kA + A'k + Q) = kBR^{-1}B'k, \quad k(1, \varepsilon) = 0.$$

(Note that Jameson and O'Malley [43] discussed the corresponding algebraic Riccati problem. A somewhat different discussion of singularly perturbed Riccati equations is contained in Womble, Potter, and Speyer [89].)

Past experience suggests that we seek a solution for the Riccati gain k in the form

(13) $$k(t, \varepsilon) = K(t, \varepsilon) + \varepsilon l(\sigma, \varepsilon)$$

where the outer solution K has a power series solution

$$K(t, \varepsilon) \sim \sum_{j=0}^{\infty} K_j(t) \varepsilon^j$$

and the boundary layer correction εl satisfies

$$l(\sigma, \varepsilon) \sim \sum_{j=0}^{\infty} l_j(\sigma) \varepsilon^j$$

with the terms all tending to zero as

$$\sigma = (1-t)/\varepsilon$$

tends to infinity. Thus K must satisfy

(14) $$\varepsilon^2 \dot{K} + \varepsilon^2(KA + A'K + Q) = KBR^{-1}B'K$$

for $t < 1$ and its limit K_0 must satisfy the reduced equation $K_0 BR^{-1} B' K_0 = 0$, so that

(15) $$B'K_0 = 0.$$

In the unlikely event that B is square and nonsingular, we have the unique solution $K_0 = 0$. Otherwise (15) merely restricts K_0 to lie in the null space of B'. Indeed (15) makes (14) a "singular" singular-perturbation problem, beyond the reach of standard techniques for singularly perturbed initial value problems (see Vasil'eva [83] or O'Malley [63]). One might also proceed by making explicit use of the matrix pseudoinverse (see, for example, Campbell [7] and Campbell, Meyer, and Rose [8]).

We shall manipulate the equation (14) for K. Postmultiplying by B provides an equation for $(KB)^{\cdot}$. Premultiplying by B' and equating coefficients of ε^2 then imply that

$$(B'K_1B)R^{-1}(B'K_1B) = B'QB > 0$$

and this allows us to solve for $B'KB \sim \varepsilon B'K_1B > 0$. Thus $(B'KB)^{-1} = O(\frac{1}{\varepsilon})$ is taken to be positive-definite and

(16) $$B'K = \varepsilon^2 R(B'KB)^{-1}[B_1'K + B'Q + (B'K)^{\cdot} + B'KA]$$

where $B_1 = AB - \dot{B}$. Backsubstituting into (14) finally yields the substitute equation

(17) $$\dot{K} + KA + A'K + Q = [KB_1 + QB + (KB)^{\cdot} + A'KB]$$
$$[B'QB + B'(KB)^{\cdot} + B'KB_1 + B'A'KB]^{-1}[B_1'K + B'Q + (B'K)^{\cdot} + B'KA].$$

Differential equations for successive terms K_j now follow by equating coefficients successively in (17).

When $\varepsilon = 0$ we obtain the parameter-free Riccati equation

(18) $$\dot{K}_0 + K_0 A_1 + A_1' K_0 + Q_1 = K_0 S_1 K_0, \quad K_0(1) = 0,$$

with

$$A_1 = A - B_1(B'QB)^{-1}B'Q,$$

$$Q_1 = Q - QB(B'QB)^{-1}B'Q,$$

and

$$S_1 = B_1(B'QB)^{-1}B_1' \geq 0.$$

(We note that this equation is well known (see Moylan and Moore [61]).) Introducing

(19) $$P_1 = I_n - B(B'QB)^{-1}B'Q,$$

we readily find that $Q_1 = P_1'QP_1 \geq 0$, so the standard linear regulator theory implies the existence of a unique positive semi-definite solution to (18).

Since $B'K_0 = 0$ must hold along solutions of (18), one might wonder whether K_0 is overdetermined. To clarify the situation, we introduce

(20) $$P_2 = I_n - P_1 = B(B'QB)^{-1}B'Q,$$

noting that P_1 and P_2 are projections ($P_i^2 = P_i$) such that

$$B'QP_1 = 0, \quad P_1 B = 0, \quad \text{and} \quad P_1 P_2 = 0.$$

Thus P_1 maps into the null space $N(B'Q)$ of $B'Q$, P_2 into the range $R(B)$ of B, P_1' into $N(B')$, and P_2' into $R(QB)$. (In the special case that P_1 is symmetric, (20) implies a direct sum decomposition of n-space.) We note that (15) implies that

(21) $$P_2'K_0 = 0$$

so the symmetric matrix K_0 satisfies

(22) $$K_0 = P_1'K_0 = K_0P_1 = P_1'K_0P_1 \,.$$

Thus (18) is actually a terminal value problem for $P_1'K_0$ and the limiting problem is not overdetermined. Because P_1 is usually singular, (18) is essentially a lower order differential equation for $P_1'K_0P_1$ in the null space of B' (see the analogous discussion of an algebraic Riccati equation by Kwatny [54]).

Higher order terms K_j satisfy linearized versions of the problem for K_0. Thus (16) implies a linear algebraic equation for $B'K_j$ (and thereby $P_2'K_j$), while (17) provides a nonhomogeneous linear differential equation for $P_1'K_j = P_1'K_jP_1$. All that needs to be prescribed termwise is a terminal value

(23) $$P_1'(1)K(1,\varepsilon)P_1(1) \sim \sum_{j=0}^{\infty} P_1'(1)K_j(1)P_1(1)\varepsilon^j$$

(the first term necessarily being the zero matrix). Splitting the problem up termwise into an algebraic equation for $P_2'K$ and a differential equation for $P_1'K$, corresponds to the frequent use of auxiliary and bifurcation equations in a complementary fashion.

Because $B'K_1B > 0$ while $k(1,\varepsilon) = 0$, the outer solution must be corrected to order $O(\varepsilon)$ in a boundary layer near $t = 1$. This suggested the representation (13). Since the system (12) is satisfied by both k and the outer solution K, (13) implies that the boundary layer correction εl must be a decaying solution of the nonlinear system

(24) $$\frac{dl}{d\sigma} = -\frac{1}{\varepsilon}\left(lBR^{-1}B'K + KBR^{-1}B'l\right) + \varepsilon(lA + A'l) - lBR^{-1}B'l$$

for $\sigma \geq 0$. In particular, l_0 must satisfy

(25) $$\frac{dl_0}{d\sigma} = -l_0B(1)R^{-1}(1)B'(1)l_0 - l_0B(1)R^{-1}(1)B'(1)K_1(1) - K_1(1)B(1)R^{-1}(1)B'(1)l_0 \,.$$

Further $B'(1)l_0(0) = -B'(1)K_1(1)$ is known in terms of

$$C_0 = R^{-\frac{1}{2}}(1)B'(1)K_1(1)B(1)R^{-\frac{1}{2}}(1) > 0$$

(see (16)). Indeed, it provides the unique decaying solution of (25),

(26) $$l_0(\sigma) = -2R^{-\frac{1}{2}}(1)B'(1)l_0(0)\left[I_n + e^{2C_0\sigma}\right]^{-1} C_0^{-1} l_0'(0)B(1)R^{-\frac{1}{2}}(1),$$

that is, $P_1'(1)l_0(\sigma)$ is determined in terms of $P_2'(1)l_0(\sigma)$ and $B'(1)l_0(\sigma)$. Further decaying terms l_j follow successively as solutions of linear equations, with the needed initial value $B'(1)l_j(0) = -B'(1)K_j(1)$ known through lower order terms of the outer expansion.

The optimal trajectory must satisfy the linear initial value problem

(27) $$\varepsilon^2 \ddot{x} = \left(\varepsilon^2 A - BR^{-1}B'(K+\varepsilon l)\right)x, \quad x(0) \text{ given.}$$

Thus $B'K_0 = 0$ implies that the corresponding reduced system will be the linear equation $BR^{-1}B'K_1X_0 = 0$, so we can expect the limiting trajectory to satisfy

(28) $$B'K_1X_0 = 0.$$

The corresponding singular arc trajectory must therefore lie in the null space of $B'K_1$ (a space of rank r since we are in Case 1). Because $x(0)$ will not generally lie on this lower dimensional manifold, an initial boundary layer correction of the state is required at $t = 0$. Another boundary layer is needed at $t = 1$, due to the nonuniform convergence of the coefficient l there. Thus we are led to seeking a trajectory of the form

(29) $$x(t, \varepsilon) = X(t, \varepsilon) + m(\tau, \varepsilon) + n(\sigma, \varepsilon)$$

for endpoint boundary layers m and n. Details of that expansion are contained in the references.

C. RELATED PROBLEMS

For the preceding cheap control problem, both the outer solution for the Riccati gain (in reverse time) and for the state were initial value problems for singular singular-perturbation problems, that is, systems of the form

(30) $$\varepsilon \dot{y} = f(t, y, \varepsilon), \quad y(0) \text{ given}, \quad 0 \leq t \leq 1,$$

where the Jacobian

$$f_y(t, y, 0)$$

is singular. We shall consider such problems for m-vectors y with infinitely differentiable coefficients under the assumption

(H) *the matrix* $f_y(t, y, 0)$ *has a constant rank* k, $0 \leq k < m$, *for all* t

and y; *its nonzero eigenvalues are all stable; and its null space is spanned by* $m - k$ *linearly independent eigenvectors.*

Then the $k = m$ problem (which can be solved using Tikhonov's results (see O'Malley [63])) suggests a solution in the form

(31) $$y(t, \varepsilon) = Y(t, \varepsilon) + \Pi(\tau, \varepsilon)$$

where the boundary layer correction Π becomes asymptotically negligible as $\tau = t/\varepsilon$ tends to infinity. Further, we can expect the limiting solution $Y_0(t)$ for $t > 0$ to be a solution of the reduced problem

(32) $$f(t, Y_0, 0) = 0 ,$$

presuming it is consistent (otherwise, we cannot expect a bounded limiting solution).

For simplicity, consider only the nearly linear problem where

(33) $$f(t, y, \varepsilon) = F(t)y + G(t) + \varepsilon h(t, y, \varepsilon) .$$

Then Hypothesis (H) guarantees the existence of a smooth orthogonal matrix E such that

(34) $$EF = \begin{bmatrix} U \\ 0 \end{bmatrix}$$

is row-reduced and of rank k for every t. (E can be readily obtained in terms of the singular value decomposition of F and numerically via Householder transformations.) Splitting E as

(35) $$E = \begin{bmatrix} E_1 \\ E_2 \end{bmatrix}$$

(after the kth row) $P = E_1' E_1$ will be a projection such that $R(P) = R(F)$ and $Q = E_2' E_2$ will be a complementary projection such that $R(Q) = N(F')$ and $S = E_1 F E_1'$ is a stable $k \times k$ matrix (see O'Malley and Flaherty [69] and O'Malley [68]).

Defining

(36) $$z = Ey = \begin{pmatrix} z_1 \\ z_2 \end{pmatrix} = \begin{pmatrix} E_1 y \\ E_2 y \end{pmatrix} ,$$

we get a new initial value problem

(37) $$\begin{cases} \varepsilon \dot{z}_1 = S z_1 + E_1 F E_2' z_2 + E_1 G + \varepsilon \tilde{h}_1(z_1, z_2, t, \varepsilon) , & z_1(0) = E_1(0) y(0) , \\ \dot{z}_2 = \frac{1}{\varepsilon} E_2 G + \tilde{h}_2(z_1, z_2, t, \varepsilon) , & z_2(0) = E_2(0) y(0) , \end{cases}$$

to which we can apply the Tikhonov theory provided

(38) $$E_2 G = 0 .$$

That holds if and only if the (formal) reduced equation $FY_0 + G = 0$ is consistent. (Like the usual procedure, we seek to transform the singular singular-perturbation problem to a regular one.)

A more natural approach is through power series. Since the representation (31) implies that the outer solution $Y(t, \varepsilon)$ must satisfy

(39) $$F(t)Y = -G(t) + \varepsilon[\dot{Y} - h(Y, t, \varepsilon)]$$

as a power series in ε, we must successively have

(40) $$\begin{cases} FY_0 = -G , \\ FY_1 = \dot{Y}_0 - h(Y_0, t, 0) , \end{cases} \text{ and so on.}$$

The first equation is, of course, the reduced equation (32). Manipulating with the projections P and Q and using the invertibility of S allows us to solve for PY_0 as a linear function of QY_0, that is,

(41) $$PY_0 = -E_1' S^{-1} E_1 \left(F(QY_0) - G \right)$$

and similarly for later PY_j's as a function of QY_j. Simultaneously, $E_2 F = 0$ implies that $QF = 0$, so consistency of the reduced and later equations requires that Q multiplied by the right hand side equals zero. Thus

(42) $$(QY_0)^{\cdot} = \dot{Q}Y_0 + Qh(Y_0, t, 0)$$

and since $Y_0 = PY_0 + QY_0$, (41) implies an initial value problem for QY_0. Using

(43) $$Q(0) Y_0(0) = Q(0) y(0)$$

uniquely implies QY_0 and, thereby, Y_0. Since $P(0)Y_0(0)$ cannot be prescribed, the need for an initial boundary layer correction for $P(0)\Pi_0(0)$ is clear. Further terms follow analogously. The combined algebraic and differential equation approach allows numerical solution of these problems in cases where the usual stiff equation routines break down. Numerical work is being done with Joseph Flaherty of Rensselaer Polytechnic Institute, and will be reported soon.

Among many generalizations of the cheap control problem, consider problems for bounded scalar controls, say $|u| \leq m$. Then one can have a saturated bang-bang control with $|u| = m$ and even an infinite number of switchings.

If we generalize (5) by considering the singular example

(44)
$$\begin{cases} \dot{x} = u \, , \quad x(0) = 1 \, , \\ J(0) = \tfrac{1}{2} \int_0^1 x^2(t)dt \, , \quad |u| \leq m \, , \end{cases}$$

we obtain the optimal solution (for $m > 1$)

(45)
$$u = \begin{cases} -m \, , & 0 \leq t < 1/m \, , \\ 0 \, , & t > 1/m \, , \end{cases}$$

corresponding to the solution $u = -\delta$ obtained for $m = \infty$. Our singular perturbation analysis indicates (but does not prove) that for singular arcs of order one, the optimal control is initially saturated before transfer to a singular arc (see Flaherty and O'Malley [24]). For Case L problems, $L > 1$, the optimal control usually switches infinitely often before reaching the singular arc. Nonetheless, for many problems, our analysis suggests how to obtain a near-optimal L-switch solution.

References

[1] Brian D.O. Anderson, John B. Moore, *Linear Optimal Control* (Prentice-Hall, Englewood Cliffs, New Jersey, 1971).

[2] A.A. Andronov, A.A. Vitt, and S.E. Khaikin, *Theory of Oscillators*, second, revised edition (Pergamon Press, Oxford; New York; Toronto, Ontario; 1966).

[3] Koichi Asatani, *Studies on Singular Perturbations of Optimal Control Systems with Applications to Nuclear Reactor Control* (Institute of Atomic Energy, Kyoto University, Kyoto, 1974).

[4] David John Bell and David H. Jacobson, *Singular Optimal Control Problems* (Mathematics in Science and Engineering, 117. Academic Press [Harcourt Brace Jovanovich], London, New York, 1975).

[5] Gilmer Blankenship and Susan Sachs, "Singularly perturbed linear stochastic ordinary differential equations", *Proceedings of the 1977 Joint Automatic Control Conference*, 1232-1237, 1977.

[6] R.S. Bucy, "Two-point boundary value problems of linear Hamiltonian systems", *SIAM J. Appl. Math.* 15 (1967), 1385-1389.

[7] Stephen L. Campbell, "Optimal control of autonomous linear processes with singular matrices in the quadratic cost functional", *SIAM J. Control Optimization* 14 (1976), 1092-1106.

[8] Stephen L. Campbell, Carl D. Meyer, Jr. and Nicholas J. Rose, "Applications of the Drazin inverse to linear systems of differential equations with singular constant coefficients", *SIAM J. Appl. Math.* 31 (1976), 411-425.

[9] Gail A. Carpenter, "A geometric approach to singular perturbation problems with applications to nerve impulse equations", *J. Differential Equations* 23 (1977), 335-367.

[10] Lamberto Cesari, "Alternative methods in nonlinear analysis", *International Conference on Differential Equations*, 95-148 (Academic Press [Harcourt Brace Jovanovich], New York, San Francisco, London, 1975).

[11] K.W. Chang, "Singular perturbations of a general boundary value problem", *SIAM J. Math. Anal.* 3 (1972), 520-526.

[12] J.H. Chow, "Preservation of controllability in linear time-invariant perturbed systems", *Internat. J. Control* 25 (1977), 697-704.

[13] J.H. Chow and P.V. Kokotović, "A decomposition of near-optimum regulators for systems with slow and fast modes", *IEEE Trans. Automatic Control* AC-21 (1976), 701-705.

[14] Joe H. Chow and Peter V. Kokotovic, "Two-time-scale feedback design of a class of nonlinear systems", *Proceedings of the* 1977 *Joint Automatic Control Conference*, 556-561, 1977.

[15] K.N. Chueh, C.C. Conley & J.A. Smoller, "Positively invariant regions for systems of nonlinear diffusion equations", *Indiana Univ. Math. J.* 26 (1977), 373-392.

[16] Earl A. Coddington and Norman Levinson, "A boundary value problem for a non-linear differential equation with a small parameter", *Proc. Amer. Math. Soc.* 3 (1952), 73-81.

[17] Earl A. Coddington and Norman Levinson, *Theory of Ordinary Differential Equations* (Internat. Series in Pure and Appl. Math. McGraw-Hill, New York, Toronto, London, 1955).

[18] Julian D. Cole, *Perturbation Methods in Applied Mathematics* (Blaisdell [Ginn & Co], Waltham, Massachusetts; Toronto; London; 1968).

[19] W.A. Coppel, "Matrix quadratic equations", *Bull. Austral. Math. Soc.* 10 (1974), 377-401.

[20] W.A. Coppel, "Linear-quadratic optimal control", *Proc. Roy. Soc. Edinburgh Sect. A* 73 (1975), 271-289.

[21] C.A. Desoer, "Singular perturbation and bounded input bounded-state stability", *Electronic Letters* 6 (1970), 16-17.

[22] Charles A. Desoer, "Distributed networks with small parasitic elements: input-output stability", *IEEE Trans. Circuits and Systems* 24 (1977), 1-8.

[23] Paul C. Fife, "Singular perturbation and wave front techniques in reaction-diffusion problems", *Asymptotic Methods and Singular Perturbations* (SIAM-AMS Proceedings, 10, 23-50. American Mathematical Society, Providence, Rhode Island, 1976).

[24] Joseph E. Flaherty and Robert E. O'Malley, Jr., "On the computation of singular controls", *IEEE Trans. Automatic Control* 22 (1977), 640-648.

[25] B. Francis & K. Glover, "Bounded peaking in the optimal linear regulator with cheap control" (Department of Engineering Report TR 151, University of Cambridge, Mass., 1977).

[26] Marvin I. Freedman and James L. Kaplan, "Perturbation analysis of an optimal control problem involving Bang-Bang controls", *J. Differential Equations* **25** (1977), 11-29.

[27] Bernard Friedland, "Limiting forms of optimum stochastic linear regulators", *Trans. ASME, Ser. G. J. Dynamic Systems, Measurement and Control* **93** (1971), 134-141.

[28] V.Ja. Glizer and M.G. Dmitriev, "Singular perturbations in a linear optimal control problem with quadratic functional", *Soviet Math. Dokl.* **16** (1975), 1555-1558.

[29] B.S. Goh, "Necessary conditions for singular extremals involving multiple control variables", *SIAM J. Control* **4** (1966), 716-731.

[30] W.M. Greenlee and R.E. Snow, "Two-timing on the half line for damped oscillation equations", *J. Math. Anal. Appl.* **51** (1975), 394-428.

[31] Patrick Habets, "The consistency problem of singularly perturbed differential equations", *An. Şti. Univ. "Al. I. Cuza" Iaşi Secţ. I a Mat. (NS)* **20** (1974), 81-92.

[32] A.H. Haddad, "Linear filtering of singularly perturbed systems", *IEEE Trans. Automatic Control* **AC-21** (1976), 515-519.

[33] Charles A. Hadlock, "Existence and dependence on a parameter of solutions of a nonlinear two point boundary value problem", *J. Differential Equations* **14** (1973), 498-517.

[34] W.A. Harris, Jr., "Singularly perturbed boundary value problems revisited", *Symposium on Ordinary Differential Equations*, Minneapolis, 1972 (Lecture Notes in Mathematics, **312**, 54-64. Springer-Verlag, Berlin, Heidelberg, New York, 1973).

[35] D.L. Hetrick, *Dynamics of Nuclear Reactors* (University of Chicago Press, Chicago, Illinois, 1971).

[36] Yu-Chi Ho, "Linear stochastic singular control problems", *J. Optimization Theory Appl.* **9** (1972), 24-31.

[37] F.C. Hoppensteadt and Willard L. Miranker, "Differential equations having rapidly changing solutions: analytic methods for weakly nonlinear systems", *J. Differential Equations* **22** (1976), 237-249.

[38] Frederick A. Howes, *Singular Perturbations and Differential Inequalities* (Memoirs Amer. Math. Soc. **168**. Amer. Math. Soc., Providence, Rhode Island, 1976).

[39] Frederick A. Howes, "Singularly perturbed boundary value problems with turning points II", *SIAM J. Math. Anal.* (to appear).

[40] Frederick A. Howes, *Boundary and Interior Layer Behavior and Their Interaction* (Memoirs Amer. Math. Soc., to appear).

[41] David H. Jacobson, Stanley B. Gershwin, and Milind M. Lele, "Computation of optimal singular controls", *IEEE Trans. Automatic Control* **AC-15** (1970), 67-73.

[42] David H. Jacobson and Jason L. Speyer, "Necessary and sufficient conditions for optimality for singular control problems: a limit approach", *J. Math. Anal. Appl.* **34** (1971), 239-266.

[43] Antony Jameson and R.E. O'Malley, Jr., "Cheap control of the time-invariant regulator", *Appl. Math. Optim.* **1** (1975), 337-354.

[44] R.E. Kalman, "Contributions to the theory of optimal control", *Bol. Soc. Mat. Mexicana* (2) **5** (1960), 102-119.

[45] R.E. Kalman, "When is a linear control system optimal?", *Trans. ASME Ser. D. J. Basic Engrg.* **86** (1964), 51-60.

[46] Tosio Kato, *Perturbation Theory for Linear Operators* (Die Grundlehren der Mathematischen Wissenschaften, **132**. Springer-Verlag, Berlin, Heidelberg, New York, 1966).

[47] P.V. Kokotović and A.H. Haddad, "Controllability and time-optimal control of systems with slow and fast modes", *IEEE Trans. Automatic Control* **AC-20** (1975), 111-113.

[48] P.V. Kokotovic, R.E. O'Malley, Jr., and P. Sannuti, "Singular perturbations and order reduction in control theory - an overview", *Automatica J. IFAC* **12** (1976), 123-132.

[49] Petar V. Kokotović and Richard A. Yackel, "Singular perturbation of linear regulators: basic theorems", *IEEE Trans. Automatic Control* **AC-17** (1972), 29-37.

[50] Vladimir Kučera, "A contribution to matrix quadratic equations", *IEEE Trans. Automatic Control* **AC-17** (1972), 344-347.

[51] Vladimír Kučera, "A review of the matrix Riccati equation", *Kybernetika (Prague)* **9** (1973), 42-61.

[52] Huibert Kwakernaak, "Asymptotic root loci of multi variable linear optimal regulators", *IEEE Trans. Automatic Control* **AC-21** (1976), 378-382.

[53] Huibert Kwakernaak and Raphael Sivan, "The maximally achievable accuracy of linear optimal regulators and linear optimal filters", *IEEE Trans. Automatic Control* **AC-17** (1972), 79-86.

[54] Harry G. Kwatny, "Minimal order observers and certain singular problems of optimal estimation and control", *IEEE Trans. Automatic Control* **AC-19** (1974), 274-276.

[55] J.J. Levin, "The asymptotic behavior of the stable initial manifolds of a system of nonlinear differential equations", *Trans. Amer. Math. Soc.* **85** (1957), 357-368.

[56] Norman Levinson, "Perturbations of discontinuous solutions of non-linear systems of differential equations", *Acta Math.* **82** (1950), 71-106.

[57] J.L. Lions, *Perturbation Singulières dans les Problèmes aux Limites et en Contrôle Optimal* (Lecture Notes in Mathematics **323**. Springer-Verlag, Berlin, Heidelberg, New York, 1973).

[58] Harold D. McIntire, "The formal asymptotic solution of a nonlinear singularly perturbed state regulator" (Pre-Thesis Monograph, Department of Mathematics, University of Arizona, Tucson, 1977).

[59] P.J. Moylan, "A note on Kalman-Bucy filters with zero measurement noise", *IEEE Trans. Automatic Control* **AC-19** (1974), 263-264.

[60] Peter J. Moylan and Brian D.O. Anderson, "Nonlinear regulator theory on an inverse optimal control problem", *IEEE Trans. Automatic Control* **AC-18** (1973), 460-465.

[61] P.J. Moylan and J.B. Moore, "Generalizations of singular optimal control theory", *Automatica J. IFAC* **7** (1971), 591-598.

[62] F.W.J. Olver, *Asymptotics and Special Functions* (Computer Science and Applied Mathematics. Academic Press [Harcourt Brace Jovanovich], New York, London, 1974).

[63] Robert E. O'Malley, Jr., *Introduction to Singular Perturbations* (Applied Mathematics and Mechanics, **14**. Academic Press [Harcourt Brace Jovanovich], New York, London, 1974).

[64] R.E. O'Malley, Jr., "Boundary layer methods for certain nonlinear singularly perturbed optimal control problems", *J. Math. Anal. Appl.* **45** (1974), 468-484.

[65] R.E. O'Malley, Jr., "On two methods of solution for a singularly perturbed linear state regulator problem", *SIAM Rev.* **17** (1975), 16-37.

[66] R.E. O'Malley, Jr., "A more direct solution of the nearly singular linear regulator problem", *SIAM J. Control Optimization* **14** (1976), 1063-1077.

[67] R.E. O'Malley, Jr., "High gain feedback systems as singular singular-perturbation problems", *Proceedings of the 1977 Joint Automatic Control Conference*, 1278-1281, 1977.

[68] R.E. O'Malley, Jr., "On singular singularly-perturbed initial value problems", *Applicable Anal.* (to appear).

[69] R.E. O'Malley, Jr. and J.E. Flaherty, "Singular singular-perturbation problems" (Lecture Notes in Mathematics, **594**, 422-436. Springer-Verlag, Berlin, Heidelberg, New York, 1977).

[70] Robert E. O'Malley, Jr., and Antony Jameson, "Singular perturbations and singular arcs - Part I", *IEEE Trans. Automatic Control* **20** (1975), 218-226.

[71] Robert E. O'Malley, Jr., and Antony Jameson, "Singular perturbations and singular arcs - Part II", *IEEE Trans. Automatic Control* **22** (1977), 328-337.

[72] Robert E. O'Malley, Jr. and Joseph B. Keller, "Loss of boundary conditions in the asymptotic solution of linear ordinary differential equations, II. Boundary value problems", *Comm. Pure Appl. Math.* **21** (1968), 263-270.

[73] R.E. O'Malley, Jr. and C.F. Kung, "On the matrix Riccati approach to a singularly perturbed regulator problem", *J. Differential Equations* **16** (1974), 413-427.

[74] R.E. O'Malley, Jr. and C.F. Kung, "The singularly perturbed linear state regulator problem II", *SIAM J. Control* **13** (1975), 327-337.

[75] Franz Rellich, *Perturbation Theory of Eigenvalue Problems* (Gordon and Breach, New York, London, Paris, 1969).

[76] Robert D. Richtmyer and K.W. Morton, *Difference Methods for Initial-Value Problems*, second edition (Interscience Tracts in Pure and Applied Mathematics, **4**. Interscience [John Wiley & Sons], New York, London, Sydney, 1967).

[77] H.M. Robbins, "A generalized Legendre-Clebsch condition for the singular cases of optimal control", *IBM J. Res. Develop.* **11** (1967), 361-372.

[78] S.I. Rubinow, *Introduction to Mathematical Biology* (Interscience [John Wiley & Sons], New York, London, Sydney, 1975).

[79] P. Sannuti, "Asymptotic series solution of singularly perturbed optimal control problems", *Automatica J. IFAC* **10** (1974), 183-194.

[80] Peddapullaiah Sannuti, and Petar V. Kokotović, "Near-optimum design of linear systems by a singular perturbation method", *IEEE Trans. Automatic Control* **AC-14** (1969), 15-22.

[81] Peddapullaiah Sannuti and Parvathareddy B. Reddy, "Asymptotic series solution of optimal systems with small time delay", *IEEE Trans. Automatic Control* **AC-18** (1973), 250-259.

[82] H.L. Turrittin, "Asymptotic expansions of solutions of systems of ordinary linear differential equations containing a parameter", *Contributions to the Theory of Nonlinear Oscillations*, Volume 2, 81-116 (Princeton University Press, Princeton, New Jersey, 1952).

[83] A.B. Vasil'eva, "Asymptotic behaviour of solutions to certain problems involving non-linear differential equations containing a small parameter multiplying the highest derivatives", *Russian Math. Surveys* **18** (1963), no. 3, 13-84.

[84] А.Б. Васильева, В.Ф. Бутузов [A.B. Vasil'eva, V.F. Butuzov], Асимптотические разложения решений сингулярно возмущенных уравнений [*Asymptotic Expansions of Solutions of Singularly Perturbed Differential Equations*] (Nauka, Moscow, 1973).

[85] M.I. Višik and L.A. Lyusternik, "Initial jump for non-linear differential equations containing a small parameter", *Soviet Math. Dokl.* **1** (1960), 749-752.

[86] Wolfgang Wasow, *Asymptotic Expansions for Ordinary Differential Equations* (Pure and Applied Mathematics, **14**. Interscience [John Wiley & Sons], New York, London, Sydney, 1965. Reprinted Kreiger, Huntington, 1976).

[87] Robert R. Wilde and Petar V. Kokotović, "Optimal open- and closed-loop control of singularly perturbed linear systems", *IEEE Trans. Automatic Control* AC-18 (1973), 616-626.

[88] Ralph A. Willoughby, *Stiff Differential Systems* (The IBM Research Symposia Series. Plenum Press, New York, London, 1974).

[89] M. Edward Womble, James E. Potter, and Jason L. Speyer, "Approximations to Riccati equations having slow and fast modes", *IEEE Trans. Automatic Control* AC-21 (1976), 846-855.

[90] W. Murray Wonham, *Linear Multivariable Control: A Geometric Approach* (Lecture Notes in Economics and Mathematical Systems, 101. Springer-Verlag, Berlin, Heidelberg, New York, 1974).

[91] Richard A. Yackel and Petar V. Kokotović, "A boundary layer method for the matrix Riccati equation", *IEEE Trans. Automatic Control* AC-18 (1973), 17-24.

[92] Kar-Keung D. Young and Petar V. Kokotovic, Vadim I. Utkin, "A singular perturbation analysis of high gain feedback systems", *Proceedings of the 1977 Joint Automatic Control Conference*, 1270-1277, 1977.

DUALITY IN OPTIMAL CONTROL

R.T. Rockafellar

For many kinds of optimization problems, convexity properties are very important, and when they are present in a thorough form they lead to an interesting kind of duality. This duality is sometimes useful in methods of computation, but it also has theoretical applications, such as in the analysis of economic models where dual variables can be interpreted as prices. The study of duality, even though it may pertain to a special subclass of problems often aids in the general development of a subject by suggesting alternative ways of looking at things.

In the classical calculus of variations, convexity and duality first enter the picture in the correspondence between Lagrangian and Hamiltonian functions and in the way this is connected with necessary conditions and the existence of solutions. Expressed in terms of the Hamiltonian, the optimality conditions for an arc x pair it with an "adjoint" arc p. The pairing carries over to problems of optimal control via the maximum principle. Duality theory in this context aims at uncovering and analyzing cases where p happens to solve a dual problem for which x is in turn the adjoint arc. But although this is the principal motivation, a number of side issues have to be explored along the way, and these suggest new approaches even to problems where duality is not at stake.

1. Implicit constraints

The effects that the aim of developing duality can have on one's point of view
___
Research sponsored by the Air Force Office of Scientific Research, Air Force Systems Command, United States Air Force, under AF-AFUSR grant number 77-0546 at the University of Washington, Seattle.

are seen immediately even in the formulation of the problem. Ordinarily, an optimal control problem for an arc x involves systems of constraints of various types. If the objective is to pass to a dual problem of similar type for an arc p, a means must be found for dualizing the constraint structure. The more details that are built into the model, the more there is to dualize, and by the time every possibility is covered in a symmetric fashion the framework may be impossibly cumbersome. It is here that the idea of representing constraints abstractly by infinite penalties has its origin.

To introduce the idea in a more elementary setting, consider first the problem of minimizing a function $F_0(z)$ over all $z \in C \subset R^N$, where F_0 is a real-valued function. The set C could be described by conditions of various kinds, for instance as the set of points satisfying equations or inequalities, but at the moment we need not be concerned with that. The point is that the problem can be represented notationally in terms of minimizing a certain *extended*-real-valued function F over the *whole* space R^N, namely

$$(1) \qquad F(z) = \begin{cases} F_0(z) & \text{if } z \in C, \\ +\infty & \text{if } z \notin C. \end{cases}$$

Indeed, if $C \neq \emptyset$ the only points of interest in minimizing F are those in C, where F agrees with F_0. The case where $C = \emptyset$ (that is the problem has no "feasible solutions") corresponds to $\min F = +\infty$.

What functions $F : R^N \to \overline{R}$ (where $\overline{R} = R \cup \{\pm\infty\}$) are of the form (1) for some *nonempty* C and real-valued F_0? They are, of course, the ones such that $F(z) > -\infty$ for all $z \in R^N$ and $F(z) < \infty$ for at least one $z \in R^N$. Such a function on R^N will be termed "*proper*".

Although topological properties of F clearly must be essential in any discussion of minimization, continuity would generally be too much to ask for, if for no other reason than because jumps to $+\infty$ are allowed at the boundary of C. A more appropriate concept is *lower semicontinuity* (l.s.c.), where the level sets of the form $\{z \in R^N \mid F(z) \leq \alpha\}$ are all required to be closed, or *inf-compactness*, where the sets in question are compact. Inf-compactness implies that F attains its minimum. Note that F is inf-compact in particular if it is of the form (1) with C compact and F_0 continuous relative to C. But F can also be l.s.c., or even inf-compact, without its effective domain $C = \{z \in R^N \mid F(z) < \infty\}$ necessarily being closed. An example in one dimension is

$$F(z) = \begin{cases} \sec z & \text{if } -\pi/2 < z < \pi/2, \\ +\infty & \text{otherwise.} \end{cases}$$

Geometrically, lower semicontinuity is equivalent to the closedness of the *epigraph* of F, which is the set

$$\text{epi } F = \{(z, \alpha) \in R^N \times R \mid \alpha \geq F(z)\}.$$

The projection of this set on R^N is C, but of course the projection of a closed set is not always closed, as the example shows.

These observations may be summarized by saying that the constrained minimization problems in R^N which are "reasonable" can be identified abstractly with the functions $F : R^N \mapsto \overline{R}$ which are proper and lower semicontinuous. The constraints are implicit in the condition $F(z) < \infty$.

2. Representation of a control example

A typical problem in optimal control might have the form: minimize

(2) $$\int_{t_0}^{t_1} f_0(t, x(t), u(t)) dt + l_0(x(t_0), x(t_1))$$

subject to

(3) $$\dot{x}(t) = f(t, x(t), u(t)), \quad u(t) \in U(t),$$
$$x(t) \in X(t), \quad (x(t_0), x(t_1)) \in E,$$

where $X(t) \subset R^n$, $U(t) \subset R^m$ and $E \subset R^n \times R^n$ (these sets may be given by explicit constraints), and x and u range over certain function spaces X and U over the fixed interval $[t_0, t_1]$. Setting aside temporarily the issue of measurability with respect to t, let us see how the problem could be represented using the idea of implicit constraints as above, but in a somewhat more subtle fashion. For (t, x, v, u) in $[t_0, t_1] \times R^n \times R^n \times R^m$, define

(4) $$K(t, x, v, u) = \begin{cases} f_0(t, x, u) & \text{if } x \in X(t), \ u \in U(t) \text{ and } f(t, x, u) = v, \\ +\infty & \text{otherwise,} \end{cases}$$

and for (x_0, x_1) in $R^n \times R^n$ define

(5) $$l(x_0, x_1) = \begin{cases} l_0(x_0, x_1) & \text{if } (x_0, x_1) \in E, \\ +\infty & \text{otherwise.} \end{cases}$$

It will be argued that the stated problem can be identified with that of minimizing the functional

(6) $$J(x, u) = \int_{t_0}^{t_1} K(t, x(t), \dot{x}(t), u(t)) dt + l(x(t_0), x(t_1))$$

over all $x \in X$ and $u \in U$. Certain conventions must, however be adopted in the definition of J.

One source of difficulty in the definition is that the expression $k(t) = K(t, x(t), \dot{x}(t), u(t))$ needs to be measurable in t, and this will be discussed below. But even if it is measurable, it might not be summable (finitely integrable) in the usual sense. Of course, if $k(t) \geq \beta(t)$ for a summable function β the integral has a well defined classical value which is either finite or $+\infty$. Likewise, if $k(t) \leq \alpha(t)$ for a summable function α the integral is either finite or $-\infty$. The only truly ambiguous case is the one where neither of these alternatives holds, and then we adopt in (6) the convention that the integral is $+\infty$ (if the need ever arises). This convention is equivalent to saying that in the formula $\int k = \int k^+ + \int k^-$, where k^+ and k^- are the positive and negative parts of k, the case $\infty - \infty$, if it occurs, should be resolved as $+\infty$. The latter rule is also the one we adopt in (6) if the integral is $-\infty$ but $l(x(t_0), x(t_1)) = +\infty$.

Under these conventions, it is clear that

(7) $$J(x, u) < \infty \Rightarrow \begin{cases} K(t, x(t), \dot{x}(t), u(t)) < \infty & \text{almost everywhere in } t, \\ l(x(t_0), x(t_1)) < \infty, \end{cases}$$

and hence the constraints (3) are satisfied (for almost every $t \in [t_0, t_1]$, still assuming measurability). Moreover $J(x, u)$ then reduces to the expression (2), so the problem is represented as claimed.

The approach we shall follow is to treat control problems in the framework of minimizing functionals of the form (6) for K and l of an appropriate general class. The interval $[t_0, t_1]$ will be fixed, but this is not an important restriction, since problems with variable time intervals can usually be recast in this form by a change of parameters. A fixed time interval is needed partly in order that the function spaces X and U over which the minimization takes place have a linear structure, as is prerequisite to the discussion of convexity. In fact, X will be taken to be the space of all absolutely continuous functions and U the space of all Lebesgue measurable functions.

3. Measurability

One of the tasks before us is to delineate a good class of functions K to use in (6). An essential property is that the Lebesgue measurability of the integrand should follow from that of $x(t)$, $\dot{x}(t)$ and $u(t)$. But to be useful, the conditions on K must be readily verifiable in terms of natural assumptions on the underlying data, for instance on f_0, f, X and U in the case of K given by (4). Furthermore, the conditions must be technically robust, in the sense of being easy to handle and preserved under the constructions and transformations that the theory will require.

Fortunately there is a simple and natural answer to the question of what conditions to impose. It has developed in recent years in close relation to the theory of measurable selections and is centered on the notion of a "normal integrand". An exposition in detail may be found in [29], and we shall limit ourselves here to quoting a few pertinent facts.

To save notation, the interval $[t_0, t_1]$ will be denoted by T. A *normal integrand* on $T \times R^N$ is a function $F : T \times R^N \mapsto \overline{R}$ such that $F(t, z)$ is lower semicontinuous in z for fixed t and measurable in (t, z) with respect to the σ-algebra generated by products of Lebesgue sets in T and Borel sets in R^N. The latter property implies in particular that $F(t, z(t))$ is Lebesgue measurable in t when $z(t)$ is. (This would be false for $F(t, z)$ merely Lebesgue measurable in (t, z). It would be true of course for $F(t, z)$ Borel measurable in (t, z), but Borel measurability turns out not to be preserved by some of the operations we will need to perform.) In particular, if $F(t, z) \equiv F_0(z)$, where F_0 is lower semicontinuous, then F is normal.

A normal integrand F is *proper* if $F(t, z)$ is a proper function of z (in the sense of §1) for every $t \in T$. Such an integrand may be construed as representing the kind of structure inherent in a "reasonable" constrained minimization problem, but with "measurable" dependence on the parameter t.

A *Carathéodory integrand* is a finite function F on $[0, T] \times R^N$ such that $F(t, z)$ is continuous in z and Lebesgue measurable in t. This is a classical notion, of which the present one may be viewed as a natural "one-sided" extension. It can be shown that F is a Carathéodory integrand if and only if both F and $-F$ are proper normal integrands. The pointwise supremum of a countable family of Carathéodory integrands is normal, although not necessarily finite or continuous everywhere.

The connection with measurable multifunctions is very important. A *multifunction*

$\Gamma : T \to R^N$ assigns to each $t \in T$ a set $\Gamma(t) \subset R^N$ (possibly empty), and it is *closed-valued* if $\Gamma(t)$ is always closed. A closed-valued multifunction is said to be *measurable* if for every closed $C \subset R^N$ the set

$$\Gamma^{-1}(C) = \{t \in T \mid \Gamma(t) \cap C \neq \emptyset\}$$

is Lebesgue measurable. If Γ is single-valued ($\Gamma(t)$ is a singleton for every t), this reduces to the usual concept for functions.

The main fact is that Γ is a closed-valued measurable multifunction if and only if it has a Castaing representation, that is the set $D = \{t \in T \mid \Gamma(t) \neq \emptyset\}$ is Lebesgue measurable and there is a countable collection $\{z_i\}_{i \in I}$ of Lebesgue measurable functions $u_i : D \mapsto R^N$ such that

$$\Gamma(t) = \mathrm{cl}\{z_i(t) \mid i \in I\} \quad \text{for every } t \in D.$$

As a corollary, one has a fundamental theorem on *measurable selections*: if $\Gamma : T \to R^N$ is closed-valued and measurable, then the set D above is Lebesgue measurable and there is a Lebesgue measurable function $z : D \to R^N$ such that $z(t) \in \Gamma(t)$ for all $t \in D$. (This is not the most general selection theorem, but it covers a vast number of applications; for a survey of selection theory, see [32].)

It happens that a function $F : T \times R^N \to \overline{R}$ is a normal integrand if and only if its epigraph multifunction

$$t \mapsto \mathrm{epi}\, F(t, \cdot) = \{(z, \alpha) \in R^{N+1} \mid \alpha \geq F(t, z)\}$$

is closed-valued and measurable. (This property is used as the *definition* of normality in the general theory where T is replaced by an arbitrary measurable space.) On the other hand, a multifunction $\Gamma : T \mapsto R^N$ is closed-valued and measurable if and only if its *indicator integrand*

(8) $$F(t, z) = \begin{cases} 0 & \text{if } z \in \Gamma(t), \\ \infty & \text{if } z \notin \Gamma(t), \end{cases}$$

is normal.

Normality has been established for all integrands of the general form

$$F(t, z) = \begin{cases} F_0(t, z) & \text{if } F_i(t, z) \leq c_i(t) \text{ for all } i \in I, \\ +\infty & \text{otherwise,} \end{cases}$$

where I is a countable (or finite or empty) index set, c_i is Lebesgue measurable, and F_0 and each F_i is a normal integrand (for example, a Carathéodory integrand).

Taking $F_0 \equiv 0$, one gets an indicator as in (8) and can conclude that a certain multifunction described by explicit constraints is measurable. For further examples and details, see [29].

4. Control model

Some basic assumptions that will remain in force may now be stated.

ASSUMPTION 1. *K is a proper normal integrand on $T \times (R^n \times R^n \times R^m)$.*

ASSUMPTION 2. *l is a proper lower semicontinuous function on $R^n \times R^n$.*

Assumption 1 implies in particular that $K(t, x(t), v(t), u(t))$ is Lebesgue measurable in t when $x(t)$, $v(t)$ and $u(t)$ are. Let A be the space of absolutely continuous functions $x : T \to R^n$, and let L be the space of Lebesgue measurable functions $u : T \to R^m$. For $x \in A$ the derivative $\dot{x}(t)$ exists almost everywhere and is Lebesgue measurable. Hence for every $x \in A$ and $u \in L$ the functional J in (6) is well-defined under the conventions for $\pm\infty$ explained in §1. The problem to be studied is

(Q)   minimize $J(x, u)$ over all $x \in A$, $u \in L$.

For this problem, (7) holds, and this means in terms of the sets

(9)
$$D(t, x, u) \triangleq \{v \in R^n \mid K(t, x, v, u) < \infty\},$$
$$U(t, x) \triangleq \{u \in R^m \mid D(t, x, u) \neq \emptyset\},$$
$$X(t) \triangleq \{x \in R^n \mid U(t, x) \neq \emptyset\},$$
$$E \triangleq \{(x_0, x_1) \in R^n \times R^n \mid l(x_0, x_1) < \infty\}$$

that one has the implicit constraints

(10)
$$\dot{x}(t) \in D(t, x(t), u(t)) \quad \text{almost everywhere,}$$
$$u(t) \in U(t, x(t)) \quad \text{almost everywhere,}$$
$$x(t) \in X(t) \quad \text{almost everywhere,}$$
$$(x(t_0), x(t_1)) \in E.$$

If these are not satisfied by any $x \in A$ and $u \in L$, then the minimum in (Q) is attained but is $+\infty$. Of course, x is interpreted as the *state trajectory* for a system being modelled, and u is the control.

In the example in §2, what assumptions suffice for the corresponding K and l to fit the conditions above for (Q)? If E is a nonempty closed set and l_0 is continuous (finite) on E, then l is certainly proper and lower semicontinuous. If the multifunctions $t \mapsto X(t)$ and $t \mapsto U(t)$ are nonempty-closed-valued and measurable, and if $f_0(t, x, u)$ and $f(t, x, u)$ are continuous in (x, u) and Lebesgue measurable in t, then K in (4) is a proper normal integrand. The latter follows from the normality criteria furnished in §3 and the elementary fact that the sum of proper normal integrands is normal. (The equation $v - f(t, x, u) = 0$ can be expressed by a finite number of constraints $f_i(t, x, v, u) \leq 0$ with f_i a Carathéodory integrand.)

The optimal control problem (Q) is said to be of *convex type* if $K(t, x, v, u)$ is convex in (x, v, u) and $l(x_0, x_1)$ is convex in (x_0, x_1). (A function $F : R^N \to \overline{R}$ is *convex* if its epigraph is a convex set, or equivalently, if the inequality $F((1-\lambda)z_0 + \lambda z_1) \leq (1-\lambda)F(z_0) + \lambda F(z_1)$ holds for all $z_0 \in R^N$, $z_1 \in R^N$ and $\lambda \in (0, 1)$ under the obvious conventions for manipulating $\pm\infty$ and, if necessary, the special rule $\infty - \infty = \infty$.) If (Q) is of convex type, then J is a convex functional on the space $A \times L$, as can easily be verified. This case will be especially important for the theory of duality.

A problem of convex type that will serve nicely to illustrate the theory at several stages is

$$(Q_0) \quad \text{minimize} \int_T f(t, C(t)x(t))dt + \int_T g(t, u(t))dt + l(x(t_0), x(t_1))$$

$$\text{subject to } \dot{x}(t) = A(t)x(t) + B(t)u(t) \text{ almost everywhere,}$$

where f and g are *convex*, proper, normal integrands (that is the functions $f(t, \cdot)$ and $g(t, \cdot)$ are convex - we are never interested in convexity with respect to t), l is convex, proper, lower semicontinuous, and the elements of the matrices $A(t)$, $B(t)$ and $C(t)$ depend Lebesgue measurably on t. This corresponds to

$$(11) \quad K(t, x, v, u) = \begin{cases} f(t, C(t)x) + g(t, u) & \text{if } v = A(t)x + B(t)u, \\ +\infty & \text{otherwise.} \end{cases}$$

It is not hard to show that K is a convex normal integrand; to ensure that K is proper, we assume for simplicity that $f(t, 0) < \infty$ for all t. The vector $y(t) = C(t)x(t)$ might be interpreted in some cases as the "observation" associated with the state $x(t)$.

One special case we shall refer to is

$$(12) \quad f(t, y) \equiv 0, \quad g(t, u) = \begin{cases} 0 & \text{if } \|u\| \leq 1, \\ \infty & \text{if } \|u\| > 1, \end{cases}$$

where $\|\cdot\|$ denotes an arbitrary norm on R^m. Then (Q_0) consists of minimizing

$l(x(t_0), x(t_1))$ subject to $\|u(t)\| \leq 1$ for almost every t and $\dot{x} = Ax + Bu$. Another case is

(13) $$f(t, y) = \tfrac{1}{2} y \cdot S(t) y, \quad g(t, u) = \tfrac{1}{2} u \cdot R(t) u,$$

where $S(t)$ and $R(t)$ are positive semidefinite matrices depending Lebesgue measurably on t. (Then f and g are Carathéodory integrands.) Note that the first integrand in (Q_0) is then $\tfrac{1}{2} x(t) \cdot Q(t) x(t)$ where $Q = C^*SC$, C^* being the transpose of C. Any positive *semidefinite* (symmetric) Q can be written in this form for some C and positive *definite* S (which are elementary to construct without resorting to eigenvectors or the like).

For the boundary function l, a simple case where it is lower semicontinuous proper convex is

(14) $$l(x_0, x_1) = \begin{cases} 0 & \text{if } x_0 = a_0, \; x_1 = a_1, \\ \infty & \text{if } x_0 \neq a_0 \text{ or } x_1 \neq a_1, \end{cases}$$

where a_0 and a_1 are two given points in R^n. This corresponds to the implicit fixed endpoint constraint $x(t_0) = a_0$, $x(t_1) = a_1$. A case involving endpoints which are not fixed, yet mutually related, is

(15) $$l(x_0, x_1) = \begin{cases} 0 & \text{if } x_0 = x_1, \\ \infty & \text{if } x_0 \neq x_1. \end{cases}$$

Then $x(t_0)$ can be arbitrary, but $x(t_1) = x(t_0)$. A mixed example is

(16) $$l(x_0, x_1) = \begin{cases} \tfrac{1}{2} |x_1 - a_1|^2 & \text{if } x_0 \in E_0, \\ \infty & \text{otherwise,} \end{cases}$$

where E_0 is a nonempty closed convex set (reducing perhaps to a single point a_0) and a_1 is a given point in R^n. Then $x(t_0)$ must lie in E_0.

5. Reduced problem

For some purposes, it is useful to know that the problem (Q) can be reduced to another form where the control u does not appear explicitly. This is a good approach in proving the existence of solutions and in drawing parallels with the classical calculus of variations. Also, much of the general duality theory applies mainly to the state trajectory $x(t)$ and an adjoint trajectory $p(t)$, although in special cases like (Q_0) it will turn out that there are natural dual controls $w(t)$ to single out for association with $p(t)$.

Starting from the fact that the optimal value in (Q) can be expressed as

(17) $\quad \inf(Q) = \inf_{x \in A} \left\{ l(x(t_0), x(t_1)) + \inf_{u \in L} \int_T K(t, x(t), \dot{x}(t), u(t)) dt \right\}$,

we are led to ask whether the minimization over $u \in L$ can be executed simply by choosing for each t a point $u(t) \in \Gamma(t)$, where

(18) $\quad \Gamma(t) = \arg \min K(t, x(t), \dot{x}(t), \cdot)$.

Of course, for this to be true the minimizing set $\Gamma(t)$ must be nonempty for almost every t, but there is also an important question of measurability. How do we know we can select $u(t) \in \Gamma(t)$ in such a way that the function u belongs to the space L? More generally, apart from whether the minimum is attained, there is the question of conditions under which the equation

(19) $\quad \inf_{u \in L} \int_T F(t, u(t)) dt = \int_T [\inf_{u \in R^n} F(t, u)] dt$

is valid, specifically when $F(t, u(t)) = K(t, x(t), \dot{x}(t), u(t))$.

It is demonstrated in [29, §3] that (19) is true for any normal integrand F, the function

$$t \mapsto \inf_{u \in R^m} F(t, u)$$

and the multifunction

$$t \mapsto \arg \min_{u \in R^m} F(t, u)$$

always being measurable. To the extent that a measurable multifunction is nonempty-valued, it has a measurable selection, as noted in §3 above. The chain of facts needed here is completed by the result in [29] that for $F(t, \cdot) = K(t, x(t), v(t), \cdot)$, the normality of F follows from that of K and the measurability of $x(t)$ and $v(t)$. (In the case of $v(t) = \dot{x}(t)$, there is a minor difficulty with the fact that $\dot{x}(t)$ may be undefined on a certain set of measure zero. This technicality can be handled by supplying an arbitrary definition over that set or by passing to a subset of T of full measure. It causes no real trouble and, for simplicity of exposition, it will be ignored wherever it crops up.)

It follows that for every $x \in A$ the functional

(20) $$\Phi(x) = \int_T L\bigl(t, x(t), \dot{x}(t)\bigr)dt + l\bigl(x(t_0), x(t_1)\bigr)$$

is well defined, where

(21) $$L(t, x, v) = \inf_{u \in R^n} K(t, x, v, u)$$

and moreover

(22) $$\inf_{u \in L} J(x, u) = \Phi(x) ,$$

where the infimum (if not $-\infty$) is attained by u if and only if u is a measurable selection (almost everywhere) for the multifunction (18). The *reduced problem* associated with (Q) is

(P) $\qquad\qquad$ minimize $\Phi(x)$ over all $x \in A$,

and L is called the *Lagrangian*. The main conclusion is thus the following

REDUCTION THEOREM. *It is always true that* $\inf(Q) = \inf(P)$. *A pair* $(x, u) \in A \times L$ *solves* (Q) *if and only if* x *solves* (P) *and* u *is a measurable selection (almost everywhere) for the multifunction* (18). *In particular, such a selection always exists if* $K(t, x, v, u)$ *is* inf-*compact in* $u \in R^m$ *for every* (t, x, v) *in* $T \times R^n \times R^n$.

This result demonstrates that one can focus all attention temporarily on x, if this is convenient, and pull the control u out of the hat at the last moment. Note that K is not uniquely determined by L, and indeed, the reduced problem (P) may arise from many different control problems (Q), corresponding to different ways of parameterizing the dynamics. In particular, any problem of the form (P) can be regarded as a problem (Q) where u does not actually appear (the control space is zero-dimensional). There is interest therefore in working directly with L, without reference to any particular K, and the basic properties assumed for L must be specified directly. It is obvious that these should be as follows.

ASSUMPTION 3. L *is a proper normal integrand on* $T \times (R^n \times R^n)$.

If L arises from a normal integrand K as in (21), then L is normal if $L(t, x, v)$ is lower semicontinuous in (x, v). This is shown by [29, Proposition 2R]. One criterion under which the proper normality of L is just a consequence of the proper normality of K is a sort of *uniform* inf-compactness of $K(t, x, v, u)$ in u: for each $t \in T$, $\alpha \in R$, and bounded set $B \subset R^n \times R^n$, the set

$$\{u \in R^m \mid \exists (x, v) \in B \text{ with } K(t, x, v, u) \leq \alpha\}$$

is bounded.

The problem (P) is said to be of *convex* type if $L(t, x, v)$ is convex in (x, v) and $l(x_0, x_1)$ is convex in (x_0, x_1). Then Φ is a convex functional on $\mathcal{A}$. The convexity of K implies the convexity of L in (21), so (P) is of convex type when (Q) is of convex type.

This holds in particular for the convex control problem (Q_0), where

(23) $\qquad L(t, x, v) = F(t, C(t)x) + \inf_u \{g(t, u) \mid B(t)u = v - A(t)x\}$.

Formula (23) uses the convention that the infimum of an empty set of real numbers is $+\infty$. The lower semicontinuity of L in (x, v) (and hence normality) follows in this case from something simpler than the "uniform inf-compactness" condition just mentioned. It suffices to have $g(t, u)$ inf-compact in u for each t.

6. Hamiltonian Function

Associated with the Lagrangian L on $T \times R^n \times R^n$ is another function H on $T \times R^n \times R^n$ which will be called the *Hamiltonian* for (P). It is defined by

(24) $\qquad H(t, x, p) = \sup_{v \in R^n} \{p \cdot v - L(t, x, v)\}$.

The Hamiltonian plays an extremely important role in many phases of variational theory, and the correspondence between Hamiltonians and Lagrangians furnishes a preliminary case of the kind of duality we aim at exploring more deeply.

Some insight into the definition of H and its classical ramifications can be gained by seeing how the formula might be applied if $L(t, x, v)$ happened to be differentiable in v. Setting the gradient of the expression to be maximized with respect to v equal to 0, one obtains the condition $p = \nabla_v L(t, x, v)$ as necessary for v to give the maximum for a particular choice of t, x and p. Suppose this can be solved for v as a function: $v = V(t, x, p)$. Then

$$H(t, x, p) = p \cdot V(t, x, p) - L(t, x, V(t, x, p)).$$

This procedure for passing from a function of v to one of p is called the *Legendre transformation*, and it is the one used in defining the Hamiltonian in the classical calculus of variations. However, it is unsatisfactory in several respects even in that framework: very strong assumptions are needed to ensure that $V(t, x, p)$ is well defined even in a local sense, and there are many technical troubles caused by the vagueness of what the true domain of H is, and the extent to which the transformation is invertible. To put this approach in a truly rigorous and suitable global form, it would be necessary to assume that $L(t, x, v)$ was not only differentiable everywhere in v, but strictly convex and subject to a certain global growth condition (coercivity). Such restrictions would be severe and, of course, would exclude most of the cases we are interested in here.

Fortunately, there is a modern alternative to the Legendre transformation which has the vigor and generality we desire. It was introduced by Fenchel [18] in 1949 and has since become a fundamental tool in convex analysis (see [20]). For any function $F : R^N \mapsto \overline{R}$, the *Fenchel transform* of F is the function $F^* : R^N \mapsto \overline{R}$ defined by

$$F^*(w) = \sup_{z \in R^N} \{w \cdot z - F(z)\} .$$

The Fenchel transform of F^* is in turn

$$F^{**}(z) = \sup_{w \in R^N} \{w \cdot z - F^*(w)\} .$$

It turns out that F^* and F^{**} are always convex and lower semicontinuous, and F^{**} is the *closed convex hull* of F in the following sense: if F majorizes at least one affine (linear-plus-a-constant) function, then the epigraph of F^{**} is the smallest closed convex set containing the epigraph of F; otherwise $F^{**} \equiv -\infty$. In fact if F is lower semicontinuous proper convex, then so is F^*, and $F^{**} = F$. The functions F and F^* are then said to be *conjugate* to each other. (It is also true that $F^{**} = F$ when $F \equiv +\infty$; then $F^* \equiv -\infty$.) One always has $F^{***} = F^*$, so F^* and F^{**} are always conjugate to each other.

Geometrically, the conjugate F^* of a lower semicontinuous proper convex function F amounts to a dual description of the epigraph of F as the intersection of a collection of nonvertical closed half-spaces in R^{N+1}.

These facts can be applied at once to the definition of the Hamiltonian. The formula expresses $H(t, x, \cdot)$ as the Fenchel transform of $L(t, x, \cdot)$. Therefore

$$\sup_{p \in R^n} \{p \cdot v - H(t, x, v)\} = \tilde{L}(t, x, v) ,$$

where $\tilde{L}$ is defined by taking the closed convex hull of $L(t, x, v)$ in v (in the special sense above) for each t, x. The Hamiltonian associated with $\tilde{L}$ is again H. The following result is then obtained from Assumption 3 and other facts of convex analysis.

HAMILTONIAN/LAGRANGIAN THEOREM. *The Hamiltonian $H(t, x, p)$ is always lower semicontinuous convex in p, and the inverse formula*

$$L(t, x, v) = \sup_{p \in R^n} \{p \cdot v - H(t, x, p)\}$$

holds if and only if the Lagrangian $L(t, x, v)$ is convex in v. In the latter case, the stronger property that L is convex in (x, v) is equivalent to H also being concave in x.

In particular, there is a *one-to-one* correspondence between Lagrangians L which are proper normal integrands, convex in the v argument, and certain functions H. Every property of such a function L is therefore dual, in principle, to some property of the associated H, and the theorem illustrates this in the case of the property of joint convexity in x and v.

When L arises from a control problem (Q) as in (21), the Hamiltonian can be expressed directly in terms of K by

(26) $$H(t, x, p) = \sup_{\substack{v \in R^n \\ u \in R^m}} \{p \cdot v - K(t, x, v, u)\} .$$

Thus for the control example in §2 the Hamiltonian is

(27) $$H(t, x, p) = \begin{cases} \sup_{u \in U(t)} \{p \cdot f(t, x, u) - f_0(t, x, u)\} & \text{if } x \in X(t) , \\ -\infty & \text{if } x \notin X(t) . \end{cases}$$

(Note the coefficient -1 for f_0. In much of the literature on optimal control, a variable coefficient p_0 is allowed, although necessary conditions are derived showing that p_0 must be constant and can be taken as either -1 or 0.) For the convex model (Q_0) where K is given by (11), the Hamiltonian is

(28) $$H(t, x, p) = p \cdot A(t)x - f(t, C(t)x) + g^*(t, B^*(t)p) \text{ with } \infty - \infty = -\infty ,$$

where $g^*(t, \cdot)$ is the convex function conjugate to $g(t, \cdot)$ for each t. (The fact that the convention $\infty - \infty = -\infty$ is needed in (28), rather than $\infty - \infty = \infty$, should serve as a warning that such conventions must be tied to specific situations and not taken for granted.)

Formulas (27) and (28) illustrate the general fact that

(29) $$H(t, x, p) = -\infty \iff x \notin X(t) ,$$

where $X(t)$ is the implicit state constraint set in (P),

(30) $$X(t) = \{x \in R^n \mid \exists v \in R^n \text{ with } L(t, x, v) < \infty\} .$$

7. Existence of Solutions

We shall come in due course to the importance of the Hamiltonian in conditions for optimality, but a few comments about its role in existence theory may now be in order. To prove the existence of a solution to (P), one needs to establish some kind of inf-compactness, or at least lower semicontinuity property of the functional Φ on the space A. Several things are involved in this, but one minimal requirement is that L should be *coercive* in v: for each (t, x), the function $L(t, x, \cdot)$ ought to be bounded below and have

$$\liminf_{|v|\to\infty} L(t, x, v)/|v| = \infty .$$

Equivalent to such coercivity is the property that for each t, x, p there should exist $\beta \in R$ such that

$$L(t, x, v) \geq p \cdot v - \beta \quad \text{for all} \quad v \in R^n .$$

But the latter inequality is equivalent by (24) to $H(t, x, p) \leq \beta$. Therefore $L(t, x, v)$ is coercive in v for each (t, x) if and only if $H(t, x, p) < \infty$ for all (t, x, p).

The classical existence theorems, such as those of Tonelli and Nagumo, require coercivity of L in v which is uniform in x. A similar requirement appears, in effect, in modern treatments of optimal control problems such as in Cesari [8], although the results are expressed in terms of a detailed constraint structure, rather than the framework of extended-real-valued Lagrangians. Matters can be kept simpler by passing to a formulation in terms of H, and in this way a broader class of existence theorems can be obtained. Olech [19] was one of the first to approach the subject from this direction, although he did not define the Hamiltonian as such.

The *Hamiltonian upper boundedness condition* is satisfied if for each $p \in R^n$ and $\beta \in R$ there is a summable function $\theta : T \mapsto R$ such that

$$H(t, x, p) < \theta(t) \quad \text{for all} \quad t \in T \quad \text{when} \quad |x| \leq \beta .$$

In particular, then H is less than $+\infty$ everywhere. To state the main consequence of this property, we need to introduce the Banach space C, consisting of all *continuous* R^n-valued functions over T, and its norm

$$\|x\|_C = \max_{t \in T} |x(t)| .$$

The space A of absolutely continuous functions is, of course, contained in C, and is a Banach space itself under the norm

(31) $$\|x\|_A = |x(0)| + \int_T |\dot{x}(t)| dt .$$

INF-COMPACTNESS THEOREM. *Suppose that the Hamiltonian upper boundedness condition is satisfied and* $L(t, x, v)$ *is convex in* v. *Then for all real numbers* α *and* β *the set*

$$\{x \in A \mid \Phi(x) \leq \alpha, \|x\|_C \leq \beta\}$$

is compact, both in the weak topology of A *and the norm topology of* C.

This is proved in [27]. It leads immediately to a result on the existence of solutions to (P) in the case where the abstract state constraint set $X(t)$ (see (30), (9), (10)) is contained for all t in a fixed bounded region of R^n. How to obtain the existence of solutions in other cases is largely a matter of finding additional growth conditions on H and l which ensure that the level sets of Φ are bounded in the norm of C, and we shall not go into it here (see [27]).

The convexity condition in the theorem deserves more elaboration, however, since it is the first place in the theory that convexity appears in an essential way, and it seems related to the Lagrangian/Hamiltonian duality. A surprising fact of functional analysis, stemming from Liapunov's theorem on the convexity of the range of a vector-valued measure is that an integral functional of the form

$$I(v) = \int_T F(t, v(t)) dt, \quad v \in L^1(T, R^n),$$

can hardly be weakly lower semicontinuous without being convex at the same time. Indeed, if one tries to take the weak closure of the epigraph of I one generally gets the epigraph of the corresponding integral functional for $F^{**}(t, \cdot)$, the convexification of $F(t, \cdot)$ described in §6 (see [29, §3] for a proof).

For functionals of the form Φ the situation is somewhat less clear, but convexity of $L(t, x, v)$ in v is crucial in much the same way. For instance, it can be shown under the Hamiltonian upper boundedness condition that any bounded sequence $\{x_k\}_{k=1}^{\infty}$ in A which is "asymptotically minimizing" for Φ (in a certain sense that will not be described here) has a subsequence converging in both the weak topology of A and the norm topology of C to an arc $x \in A$ which minimizes, not Φ, but the corresponding problem with L replaced by its convexification $\tilde{L}$ in the v argument (as defined in §6). This is called the *relaxed problem* $(\tilde{P})$, and $\tilde{L}$ is the *relaxed Lagrangian*.

The meaning of these facts is that, without the convexity of L in v, there is little motivation for studying (P), since it is likely to amount to a problem of minimizing something not possessed of a reasonable continuity property. One should look instead at $(\tilde{P})$ and its interpretation in whatever application may be at hand, since even from a computational point of view the best one could usually hope for is to generate a sequence $\{x_k\}_{k=1}^{\infty}$ converging to a solution to $(\tilde{P})$.

Other facts lend their weight to this point of view. For instance, the Weierstrass necessary condition for optimality in classical problems comes close to saying that a solution to (P) must be a solution to $(\tilde{P})$ along which the two Lagrangians L and $\tilde{L}$ happen to agree. Results of the latter sort have in fact been established for problems of optimal control under certain conditions; *cf.* Clarke [9], Warga [33].

Much can be said, therefore, in favor of compartmentalizing the theory into the study of (P) under the assumption of convexity in v on the one hand, and the study of the relationship between (P) and $(\tilde{P})$ without the assumption on the other. The second part, called *relaxation theory*, encompasses such important topics as "bang-bang" controls, as well as facts of the sort already mentioned. Whatever the merits of this philosophy, we shall follow it here in looking henceforth only at problems which are already "relaxed".

ASSUMPTION 4. $L(t, x, v)$ *is convex in* v *for every* t, x, *or in other words*, $L = \tilde{L}$.

Of course, in the main case we shall be concerned with, L will actually be convex jointly in x and v. But Assumption 4 will facilitate comparisons and conjectures having to do with more general problems.

8. Optimality conditions

One of the classical conditions for optimality of x in (P), whose necessity can be proved under certain assumptions when L and l are differentiable, is the Euler-Lagrange equation

$$\frac{d}{dt}\left[\nabla_v L(t, x(t), \dot{x}(t))\right] = \nabla_x L(t, x(t), \dot{x}(t)) .$$

This can also be expressed by asserting that for a certain function $p(t)$ one has

$$(\dot{p}(t), p(t)) = \nabla L(t, x(t), \dot{x}(t)) ,$$

where ∇L denotes the gradient of L with respect to (x, v). (As a general notational rule, we ignore t in the symbolism for gradients, conjugates, and so on, of integrands.) The corresponding condition for endpoints has the form

(32) $$(p(t_0), -p(t_1)) = \nabla l(x(t_0), x(t_1)) .$$

The key to generalizing such equations to the nondifferentiable case dictated by the present model is an appropriate substitute for the notion of "gradient".

Such a notion is well known in the case of *convex* functions. If F is convex on R^N, the *subgradient* set $\partial F(z)$ is defined to consist of all $w \in R^N$ with the property that

(33) $$F(z') \geq F(z) + w \cdot (z'-z) \quad \text{for all} \quad z' \in R^N .$$

If $F(z)$ is finite, this means that the graph of the affine function of z' on the right side of (33) is a supporting hyperplane to the epigraph of F at $(z, F(z))$. (If $F(z) = -\infty$, or if $F \equiv +\infty$, the condition is satisfied by every w, but if $F(z) = +\infty$ and $F \not\equiv +\infty$, it is not satisfied by any w.)

The theory of subgradients is presented in [20], and only a few basic facts will be cited here. The set $\partial F(z)$ is always closed and convex (possibly empty), and it reduces to a *single* element w if and only if F is differentiable at z (in which event $w = \nabla F(z)$). In the case of a lower semicontinuous proper convex function and its conjugate, satisfying

$$(34) \qquad F(z) + F^*(w) \geq z \cdot w \quad \text{for all} \quad z, w,$$

by the definition of conjugacy, there is the important, symmetric equivalence

$$(35) \qquad w \in \partial F(z) \iff F(z) + F^*(w) = z \cdot w \iff z \in \partial F^*(w).$$

A special case worthy of note is the *indicator* of a nonempty closed convex set C:

$$(36) \qquad F(z) = \begin{cases} 0 & \text{if } z \in C, \\ \infty & \text{if } z \notin C. \end{cases}$$

Then

$$(37) \qquad \partial F(z) = N_C(z) = \text{normal cone to } C \text{ at } z,$$

where

$$(38) \qquad N_C(z) = \begin{cases} \{w \in R^N \mid w \cdot (z'-z) \leq 0 \text{ for all } z' \in C\} & \text{if } z \in C, \\ \emptyset & \text{if } z \notin C. \end{cases}$$

For problems of convex type, we can work with the subgradient sets $\partial L(t, x, v)$ and $\partial l(x_0, x_1)$ in $R^N \times R^N$. The *Euler-Lagrange condition* is then

$$(39) \qquad (\dot{p}(t), p(t)) \in \partial L(t, x(t), \dot{x}(t)), \text{ almost everywhere,}$$

and the *transversality condition* is

$$(40) \qquad (p(t_0), -p(t_1)) \in \partial l(x(t_0), x(t_1)).$$

We are interested in the functions $x \in A$ which satisfy these for some $p \in A$, which is then said to be *adjoint* to x. (The adjoint arc is not necessarily unique.)

Just what these conditions, first introduced in [21], have to do with optimality in the problem (P) will be the subject of much discussion below. Before getting into that, however, we would like to mention that the definition of $\partial F(z)$ has been extended by Clarke [10], [16], to the case of arbitrary proper lower semicontinuous functions F in such a way as to coincide with the set above when F is convex and with the singleton $\{\nabla F(z)\}$ at points where F is strongly differentiable (not necessarily convex). Moreover $\partial F(z)$ is still always a closed convex set. The Euler-Lagrange condition (39) and transversality condition (40) are therefore well-defined for (P) even without any convexity assumptions. Indeed, Clarke has shown they are necessary for optimality in a number of cases [11], [13]. This more general theory falls outside of our target area of duality and will therefore not be outlined here.

Our discussion of necessity and sufficiency for optimality will be limited mainly to the convex case, where there is a reversal of the situation often encountered in variational theory: the sufficiency is the easy part.

SUFFICIENCY THEOREM. *If* (P) *is of convex type and* $x \in A$ *satisfies the Euler-Lagrange condition for* L *and transversality condition for* l *with adjoint* $p \in A$, *then* x *furnishes the minimum in* (P).

The argument is so short and simple it will be given in full. Suppose (39) and (40) hold, and let x' be an arbitrary element of A (the prime has nothing to do with derivatives). From the definition of subgradients, we have

$$L(t, x'(t), \dot{x}'(t)) \geq L(t, x(t), \dot{x}(t)) + \dot{p}(t)(x'(t)-x(t)) + p(t)(\dot{x}'(t)-\dot{x}(t))$$

for almost every t and

$$l(x'(t_0), x'(t_1)) \geq l(x(t_0), x(t_1)) + p(t_0)(x'(t_0)-x(t_0)) - p(t_1)(x'(t_1)-x(t_1)).$$

Integrating the first inequality over $[t_0, t_1] = T$ and adding the second, we obtain

$$\Phi(x') \geq \Phi(x) + \int_{t_0}^{t_1} \frac{d}{dt}[p \cdot (x'-x)]dt - [p \cdot (x'-x)]_{t_0}^{t_1},$$

where the terms in $p \cdot (x'-x)$ cancel each other.

The necessity of the conditions requires stronger assumptions, as we shall see in §13, and certain extensions have to be made in order to handle the case where the state constraint $x(t) \in X(t)$ becomes effective.

For the moment we turn instead to the question of what the conditions mean for specific cases, such as the control problem (Q_0) in §4. One thing of great practical importance in this respect is that quite a "calculus" exists for determining the subgradients of convex functions which, like L and l, are likely to be given in terms of various other functions, sets, constraints, operations, and so on (see [20], [26]).

Suppose L comes via (21) from a function $K(t, x, v, u)$ which is convex in (x, v, u). It is known that then

(41) $[(r, p) \in \partial L(t, x, v)$ and $u \in \arg\min K(t, x, v, \cdot)]$
$$\iff (r, p, 0) \in \partial K(t, x, v, u)$$

(*cf.* [26, Theorem 24 (a)]). Now suppose further that K has the form (11), so that L is given by (23), and that $g(t, \cdot)$ is inf-compact for each t, so that L is normal (as noted at the end of §5). The "arg min" set is then always nonempty, so the calculation of ∂L is reduced by (41) to that of ∂K. Assuming for each t that $f(t, \cdot)$ is finite on a neighborhood of 0 (so as to handle the case where the range space for $C(t)$ might not be all of R^n), one can show by the subgradient calculus that $(r, p, 0) \in \partial K(t, x, v, u)$ if and only if

$$v = A(t)x + B(t)u \text{ and } B^*(t)p \in \partial g(t, u),$$
$$\exists w \in \partial f(t, C(t)x) \text{ with } r = -A^*(t)p + C^*(t)w,$$

where the asterisk denotes the transpose of a matrix. Using (34), one can write the condition $B^*(t)p \in \partial g(t, u)$ in the dual form $u \in \partial g^*(t, B^*(t)p)$, where g^* is the conjugate integrand.

An application of facts about measurable selections [29] then leads to the conclusion that $x \in A$ and $p \in A$ satisfy the Euler-Lagrange condition for L in this case if and only if there exist functions $u \in L$ and $w \in L$ such that (for almost every t)

(42)
$$\dot{x}(t) = A(t)x(t) + B(t)u(t) \text{ with } u(t) \in \partial g^*(t, B^*(t)p(t)),$$
$$\dot{p}(t) = -A^*(t)p(t) + C^*(t)w(t) \text{ with } w(t) \in \partial f(t, C(t)x(t)).$$

This is interesting because of the appearance of a dual dynamical system with explicit controls $w(t)$, a property that is not readily captured for general convex K, and because of the complete symmetry in x and p. A dual problem of optimal control in p will be described in the next section.

If f and g have the quadratic form in (13), the control conditions in (42) take the form

$$w(t) = S(t)C(t)x(t) \text{ and } u(t) = R(t)^{-1}B^*(t)p(t).$$

In the case of (12), they become

$$w(t) \equiv 0 \text{ and } u(t) \in \arg\max_{\|z\| \leq 1} p(t) \cdot z.$$

If l is differentiable, the transversality condition is just (32). In the fixed endpoint case where l is given by (14), one has

$$\partial l(x_0, x_1) = \begin{cases} R^n \times R^n & \text{if } (x_0, x_1) = (a_0, a_1), \\ \emptyset & \text{if } (x_0, x_1) \neq (a_0, a_1), \end{cases}$$

so the condition reduces merely to the constraints $x(t_0) = a_0$ and $x(t_1) = a_1$, with nothing required of $p(t_0)$ and $p(t_1)$. For (15) it becomes

$$x(t_0) = x(t_1) \text{ and } p(t_0) = p(t_1),$$

while for (16) one gets

$$p(t_0) \text{ normal to } E_0 \text{ at } x(t_0), \quad -p(t_1) = Wx(t_1) + c.$$

("Normal" means "belonging to the normal cone" N_{E_0} defined in (38).)

These examples illustrate that a wide spectrum of conditions is covered by the subgradient notation. A similar calculus exists for generalized gradients in the sense of Clarke [10], [16], but it typically involves chains of inclusions rather than equivalences. Fortunately the inclusions are in the direction one needs for the derivation of necessary conditions for optimality.

9. Dual problem

The equivalent ways of writing a subgradient relation in terms of a convex function or its conjugate, as in (35), suggest a dual form for the optimality conditions for problems of convex type:

(43)
$$(r, p) \in \partial L(t, x, v) \iff (x, v) \in \partial L^*(t, r, p),$$
$$(p_0, -p_1) \in \partial l(x_0, x_1) \iff (x_0, x_1) \in \partial l^*(p_0, -p_1).$$

Here the conjugate functions $L^*(t, \cdot, \cdot)$ and l^*, like $L(t, \cdot, \cdot)$ and l, are lower semicontinuous proper convex, and in fact L^* is again a *normal integrand* [21], [29] (something which might not have been true if a different measurability property had been incorporated in the definition of "normality").

Symmetry is not quite present in (43), so let us introduce the functions

(44)
$$M(t, p, r) = L^*(t, r, p) = \sup_{x,v} \{r \cdot x + p \cdot v - L(t, x, v)\},$$
$$m(p_0, p_1) = l^*(p_0, -p_1) = \sup_{x_0, x_1} \{p_0 \cdot x_0 - p_1 \cdot x_1 - l(x_0, x_1)\},$$

so that reciprocally

(45)
$$L(t, x, v) = M^*(t, v, x) = \sup_{p,r} \{r \cdot x + p \cdot v - M(t, p, r)\},$$
$$l(x_0, x_1) = m^*(x_0, -x_1) = \sup_{p_0, p_1} \{p_0 \cdot x_0 - p_1 \cdot x_1 - m(p_0, p_1)\},$$

and the equivalences (43) become

$$(r, p) \in \partial L(t, x, v) \iff (v, x) \in \partial M(t, p, r),$$
$$(p_0, -p_1) \in \partial l(x_0, x_1) \iff (x_0, -x_1) \in \partial m(p_0, p_1).$$

For arcs $x \in A$ and $p \in A$, one therefore has

$$(\dot{p}(t), p(t)) \in \partial L(t, x(t), \dot{x}(t)) \iff (\dot{x}(t), x(t)) \in \partial M(t, p(t), \dot{p}(t)),$$
$$(p(t_0), -p(t_1)) \in \partial l(x(t_0), x(t_1)) \iff (x(t_0), -x(t_1)) \in \partial m(p(t_0), p(t_1)).$$

It is appropriate to call M the *dual Lagrangian* and m the *dual boundary function*. They satisfy the same conditions as do L and l for problems of convex type. Thus the functional

$$\Psi(p) = \int_T M(t, p(t), \dot{p}(t)) dt + m(p(t_0), p(t_1))$$

is likewise well defined for all $p \in A$ and convex. The problem

(P*) minimize $\Psi(p)$ over all $p \in A$

is the *dual* of (P) and is again of convex type. The theorem in §8 is therefore applicable and says that p solves (P*) if p satisfies the Euler-Lagrange condition for M and transversality condition for m in terms of some $x \in A$. An interesting connection between (P) and (P*) is then apparent from (46).

DUALITY THEOREM 1. *When* (P) *is of convex type, the following are equivalent for* $x \in A$ *and* $p \in A$:

(a) x *satisfies the Euler-Lagrange condition for* L *and transversality condition for* l *with adjoint* p;

(b) p *satisfies the Euler-Lagrange condition for* M *and transversality condition for* m *with adjoint* x.

Thus the sufficient conditions for (P) *also furnish a solution to* (P*) *and conversely.*

Because of the equivalence of (a) and (b) we shall simply say in the convex case that x and p satisfy the (sufficient) *optimality conditions* when these properties are present.

In the case where (P) is the reduced problem for the convex control problem (Q_0),

(46)
$$\text{minimize} \int_T f(t, C(t)x(t))dt + \int_T g(t, u(t))dt + l(x(t_0), x(t_1))$$
$$\text{subject to } \dot{x}(t) = A(t)x(t) + B(t)u(t),$$

the dual has a similar structure. Assume, as was done in §7 in specializing the optimality conditions to this setting, that for each t,

(47)
$$f(t, \cdot) \text{ is finite on a neighborhood of } 0,$$
$$g(t, \cdot) \text{ is inf-compact.}$$

These two properties happen to be dual to each other with respect to conjugate convex functions [20, §§8, 13], so (47) is equivalent to:

(48)
$$f^*(t, \cdot) \text{ is inf-compact,}$$
$$g^*(t, \cdot) \text{ is finite on a neighborhood of } 0.$$

When the expression

$$L(t, x, v) = f(t, C(t)x) + \min_u \{g(t, u) \mid B(t)u = v - A(t)x\}$$

is inserted in (44), one obtains with the help of one of the standard formulas for conjugates (*cf.* [20, p. 142]) that

(49) $\quad M(t, p, r) = g^*(t, B^*(t)p) + \min_w \{f^*(t, w) \mid C^*(t)w = r + A^*(t)p\}$.

Thus (P*) is the reduced problem for a certain control problem like (Q_0):

(50)
$$\text{minimize} \int_T g^*(t, B^*(t)p(t)) dt + \int_T f^*(t, w(t)) dt + m(p(t_0), p(t_1)),$$

$$\text{subject to } \dot{p}(t) = -A^*(t)p(t) + C^*(t)w(t).$$

Note that the dual dynamical system is the same one seen earlier in the optimality conditions (42).

Conjugate functions are not always easy to express in a more direct form, even with the machinery in [20] and [26, §9], but this is possible in many important cases. For example, if f and g have the quadratic form in (13) with $S(t)$ and $R(t)$ positive definite, one has

$$f^*(t, w) = \tfrac{1}{2} w \cdot S(t)^{-1} w, \quad g^*(t, q) = \tfrac{1}{2} q \cdot R(t)^{-1} q.$$

If they have the form (12), then

$$f^*(t, w) = \begin{cases} 0 & \text{if } w = 0, \\ \infty & \text{if } w \neq 0, \end{cases} \quad g^*(t, q) = \|q\|_*,$$

where $\|\cdot\|_*$ is the norm dual to $\|\cdot\|$. Then $w(t)$ is implicitly constrained to vanish in (50), and everything about it drops out of the problem. The same would be true in other problems with $f(t, y) \equiv 0$. Thus for a problem of the form

$$\text{minimize} \int_T g(t, u(t)) dt + l(x(t_0), x(t_1)),$$

$$\text{subject to } \dot{x}(t) = A(t)x(t) + B(t)u(t),$$

the dual is

$$\text{minimize} \int_T g^*(t, B^*(t)p(t)) dt + m(p(t_0), p(t_1)),$$

$$\text{subject to } \dot{p}(t) = -A^*(t)p(t).$$

What is particularly interesting about this case is that the dual problem turns out to be essentially *finite-dimensional*, since p is uniquely determined by $p(t_0)$.

Another good illustration is the case where

(51) $\quad f(t, y) = \begin{cases} \alpha|y| - \alpha & \text{if } |y| \geq 1, \\ 0 & \text{if } |y| \leq 1, \end{cases} \quad g(t, u) = \max\{a_1 \cdot u, \ldots, a_N \cdot u\},$

where $|\cdot|$ denotes the Euclidean norm and $a_1, \ldots, a_N$ are vectors in R^n, $\alpha > 0$. Then

$$\text{(52)} \quad f^*(t, w) = \begin{cases} |w| & \text{if } |w| \leq \alpha, \\ \infty & \text{if } |w| > \alpha, \end{cases} \quad g^*(t, q) = \begin{cases} 0 & \text{if } q \in \text{co}\{a_1, \ldots, a_N\}, \\ \infty & \text{otherwise}, \end{cases}$$

where "co" denotes convex hull. This is instructive because the primal problem (46) has no implicit state constraints or control constraints, but the dual problem (50) does, namely

$$\text{(53)} \quad B^*(t)p(t) \in \text{co}\{a_1, \ldots, a_N\} \quad \text{and} \quad |w(t)| \leq 1, \text{ almost everywhere.}$$

These constraints are determined simply by inspecting where the functions in the dual problem are finite, which underscores the economy and effectiveness of the $+\infty$ notation.

When it comes to the possibilities for l and m, the first example to look at is the one for fixed endpoints $x(t_0) = a_0$, $x(t_1) = a_1$, where l is given by (14). Trivially, m is then linear:

$$m(p(t_0), p(t_1)) = p(t_0) \cdot a_0 - p(t_1) \cdot a_1.$$

Since m is finite everywhere, no implicit constraints are imposed on the endpoints of p; they are free in the dual problem. If instead l has the form (15), corresponding to the constraint $x(t_0) = x(t_1)$, it turns out that $m = l$, so that the dual problem likewise has the constraint $p(t_0) = p(t_1)$. The example of l in (16) yields

$$\text{(54)} \quad m(p(t_0), p(t_1)) = \sigma(p(t_0)) + \tfrac{1}{2}|p(t_1) - a_1|^2 + \tfrac{1}{2}|a_1|^2,$$

where σ is the *support function* of the convex set E_0:

$$\sigma(p_0) = \sup\{p_0 \cdot x_0 \mid x_0 \in E_0\}.$$

If E_0 is a cone, σ is just the indicator of the polar cone E_0^*, and the first term in (54) represents the constraint $p(t_0) \in E_0^*$. For instance, if $E_0 = R^n$ ($x(t_0)$ free) one gets $E_0^* = \{0\}$ and the implicit constraint $p(t_0) = 0$. If $E_0 = R_+^n$ ($x(t_0) \geq 0$), then $E_0^* = R_-^n$ ($p(t_0) \leq 0$). If E_0 equals a subspace N, then $E_0^* = N^\perp$ (orthogonal complement). If E_0 is the unit ball for a norm $\|\cdot\|$, then $\sigma = \|\cdot\|_*$ (dual norm).

Incidentally, the kind of duality seen in (46) and (50), where explicit controls appear in both problems, can be captured in a slightly broader setting with the expression $f(t, C(t)x(t)) + g(t, u(t))$ replaced by $h(t, C(t)x(t), u(t))$. This replaces $f^*(t, w(t)) + g^*(t, B^*(t)p(t))$ in (50) by $h^*(t, w(t), B^*(t)p(t))$.

The dual problem (P*) was introduced in [21].

10. Hamiltonian equations

The classical reason for introducing the Hamiltonian function is that the Euler-Lagrange condition for L can, under certain assumptions, be written instead in the form

(55) $$\left(-\dot{p}(t),\, \dot{x}(t)\right) = \nabla H\left(t,\, x(t),\, p(t)\right) .$$

The same thing can be accomplished in the convex case in terms of subgradients instead of gradients.

Since for problems of convex type $H(t, x, p)$ is concave in x (as well as convex in p), we can speak of the subgradient set $\partial_p H(t, x, p)$ and, with a change of sign, the "supergradient" set $\partial_x H(t, x, p)$. The subgradient set of the function $H(t, \cdot, \cdot)$ at (x, p) is

(56) $$\partial H(t, x, p) = \partial_x H(t, x, p) \times \partial_p H(t, x, p) .$$

The generalized *Hamiltonian equation* (really: Hamiltonian "contingent equation" or "differential inclusion") is

(57) $$\left(-\dot{p}(t),\, \dot{x}(t)\right) \in \partial H\left(t,\, x(t),\, p(t)\right) , \text{ almost everywhere.}$$

The product form in (56) may give a misleading impression, in that it is a special feature which does not carry over to other classes of functions H when the definition of the Hamiltonian equation is extended. An extension is indeed possible, for example to all problems satisfying the Hamiltonian upper boundedness condition in §7. Then for each t the function

$$F : (x, p) \mapsto -H(t, x, p)$$

is lower semicontinuous proper [27, Proposition 4], so that ∂F is well defined in the sense of Clarke [10]: take

$$\partial H(t, x, p) = -\partial[-H](t, x, p) .$$

This definition turns out to give the same result as the one above if H is concave-convex, so (56) is natural for that case. But (56) is often false, although $\partial H(t, x, p)$ is always a closed convex subset of $R^n \times R^n$. (Incidentally, there are problems of convex type for which neither $H(t, x, p)$ nor $-H(t, x, p)$ is a lower semicontinuous proper function of (x, p); *cf.* [20, §33]. No general definition of ∂H is presently known which covers this case in convex analysis, having significant consequences below, and all the cases amenable to Clarke's definition.)

THEOREM. *In the convex case, the Hamiltonian equation is equivalent to the Euler-Lagrange condition for L (and also the one for M) and therefore can be substituted for it in the optimality conditions.*

This follows from a rule relating subgradients and the Fenchel transform [20, Theorem 37.5] which in the present notation takes the form

(58) $\qquad (r, p) \in \partial L(t, x, v) \iff (-r, v) \in \partial H(t, x, p)$.

Thus

(59) $\qquad (\dot{p}(t), p(t)) \in \partial L(t, x(t), \dot{x}(t)) \iff (-\dot{p}(t), x(t)) \in \partial H(t, x(t), p(t))$.

The equivalence also holds in the classical, continuously differentiable case, if L is actually strictly convex and coercive in v (not necessarily convex in (x, v)), or if H is convex in (x, p) (not necessarily differentiable), in which event L is concave in x - a reversal of the properties in the theorem in §6. But it can fail for some of the general cases covered in terms of Clarke's definition. (Then the two conditions in (59) seem to say different things, yet Clarke has established that they are both sometimes necessary for optimality. See [12], [14] for Clarke's necessary conditions in Hamiltonian form and [15] for their applications to get an extremely general "maximum principle".)

It may be wondered why in the convex case, as in the theorem above, equal attention is not paid to the *dual* Hamiltonian H' corresponding to the dual Lagrangian M,

(60) $\qquad H'(t, p, x) = \sup_{r \in R^n} \{r \cdot x - M(t, p, r)\}$.

The reason is that

$$H'(t, p, x) = -H(t, x, p) \quad \text{"almost"}.$$

Indeed, if the formula for H in terms of L is used to rewrite the formula for M in terms of L, one obtains

(61) $\qquad M(t, p, r) = \sup_{x \in R^n} \{r \cdot x + H(t, x, p)\}$,

which says that the Fenchel transform of $F(x) = -H(t, x, p)$ is $F^*(r) = M(t, p, r)$. Then from (60) one has $F^{**}(x) = H'(t, x, p)$, so the study of the relationship between H' and H boils down to the question of the extent to which F^{**} must agree with F. Since $F^{**} = F$ when F is lower semicontinuous and nowhere $-\infty$, we may conclude that $H'(t, p, x) = -H(t, x, p)$ for all (t, x, p) when H is upper semicontinuous in x and nowhere $+\infty$. Actually, for H arising from L which is lower semicontinuous proper convex in (x, v) as here, it can be shown that H is upper semicontinuous in (x, p) if it is nowhere $+\infty$. In general, however, there could be slight discrepancies between H' and H, and what one really has is two concave-convex functions equivalent to each other in a sense known in convex analysis (*cf.* [20, §34]). The Hamiltonian equations for H' and H are equivalent.

The Hamiltonian for the control problem (Q_0), expressed in (28), yields (under (47) or equivalently (48)) the equations

$\dot{x}(t) \in [A(t)x(t)+B(t)\partial g^*(t, B^*(t)p(t))]$, almost everywhere,

$\dot{p}(t) \in [-A^*(t)p(t)+C^*(t)\partial f(t, C(t)x(t))]$, almost everywhere,

which can be expanded to (42) through an application of the theory of measurable selections.

The Hamiltonian for the nonlinear control problem in §2, given in (27), may well fail to be concave-convex, yet this is a case where under natural assumptions the Hamiltonian equation is well defined in Clarke's sense. It is interesting to see how the equation relates to the maximum principle. For simplicity and in order to ensure that the reduced Lagrangian $L(t, x, v)$ is a proper normal integrand which is convex in v, as we have been assuming, suppose that

(a) $U(t)$ is compact, convex, nonempty,

(b) $f(t, x, u)$ and $f_0(t, x, u)$ are defined on all of $T \times R^n \times R^m$, measurable in t and differentiable in x,

(c) $f, f_0, \nabla_x f$ and $\nabla_x f_0$ are continuous in (x, u),

(d) f is affine in u (that is, $f(t, x, u) = F(t, x) + G(t, x)u$) and f_0 is convex in u,

(e) $x \in \text{int } X(t)$.

These conditions can be shown to imply

$$(s, v) \in \partial H(t, x, p) \iff \begin{cases} \exists u \in \arg\max_{U(t)} \{f(t, x, \cdot) \cdot p - f_0(t, x, \cdot)\} \\ \text{such that } v = f(t, x, u), \\ s = \nabla_x f(t, x, u)p - \nabla_x f_0(t, x, u). \end{cases}$$

With the help of measurable selections, this yields the result that, for $x \in A$ and $p \in A$ with $x(t) \in \text{int } X(t)$ for all t, the Hamiltonian equation is satisfied if and only if there is a measurable function u such that for almost every t,

$$u(t) \in \arg\max_{U(t)} \{f(t, x(t), \cdot) \cdot p(t) - f_0(t, x(t), \cdot)\},$$

$$\dot{x}(t) = f(t, x(t), u(t)),$$

$$\dot{p}(t) = -\nabla_x f(t, x(t), u(t))p(t) + \nabla_x f_0(t, x(t), u(t)).$$

This amounts to the "maximum principle" in reduced form. (The case where $x(t)$ might be on the boundary of $X(t)$ is more complicated, see the remarks at the end of §14.)

Note that the coefficient of f_0 is -1 in the "arg max", in contrast to most treatments of optimal control, which allow a variable coefficient $p_0(t)$ and show that it must be constant and can be taken as either -1 or 0. Since the "0" possibility is excluded, the conditions are slightly stronger than usual and require for their necessity slightly stronger assumptions (Clarke's concept of "calmness", *cf.* [13], [14]).

11. Hamiltonian trajectories

The advantage of the Hamiltonian equation over the Euler-Lagrange condition is that it has the form of a generalized ordinary differential equation

(62) $\quad \dot{z}(t) \in C(t, z(t)) \quad$ almost everywhere, $\quad z(t) = (x(t), p(t))$,

where $C(t, z)$ is a closed convex set that depends on t and z in a nice way. (The graph $\Gamma(t)$ of the multifunction $z \mapsto C(t, z)$ is closed, and the multifunction $t \mapsto \Gamma(t)$ is measurable.) Local existence theorems are available for such generalized differential equations, at least under certain conditions of nonemptiness and boundedness of $C(t, z)$ (*cf.* [7]). They can be applied to get trajectories $(x(t), p(t))$ for the Hamilton equations that emanate from any initial point (x_0, p_0) in a neighborhood of which H is Lipschitz continuous with respect to x, p , and satisfies a summability condition in t (*cf.* [22] for the convex case).

When H is concave-convex, H is not only Lipschitz continuous on any open set where it is finite, but actually differentiable there almost everywhere [20, §35], so that $\partial H(t, x, p)$ reduces to a single element (the gradient) except on a special set of measure zero. Then the general Hamiltonian equation (57) is not so far from the classical version (55) as might have been thought from its "contingent" form. As a matter of fact, nonuniqueness of solutions from a given starting point appears, from examples, to be a rather rare phenomenon, although it definitely can occur (see below).

Another property known in the convex case is that if H is finite and independent of t , then $H(x(t), p(t))$ is constant along all solutions to the generalized Hamiltonian equation. (This extends a classical result in the differentiable case whose proof is trivial, but the multivalued form of the equation requires a somewhat tricky argument, *cf.* [22].) A nice way of generating simple non-classical examples is thereby provided: take any finite concave-convex function H on $R \times R$ and look at its level curves. The trajectories of the Hamiltonian equation (which exist at least locally for this case, as just remarked) must follow these curves. A rather interesting example to look at in such a light is

$$H(x, p) = \max\{0, |p|-1\} - \max\{0, |x|-1\} ,$$

which corresponds to

$$L(x, v) = \max\{0, |x|-1\} + \begin{cases} |v| & \text{if } |v| \leq 1 , \\ \infty & \text{if } |v| > 1 . \end{cases}$$

The trajectories have corners, and they can branch at certain points.

The assertions about $\partial H(t, x, p)$ being a singleton almost everywhere, and H being constant along Hamiltonian trajectories when H is independent of t, carry over to other cases, for instance the Hamiltonian at the end of the preceding section and all Hamiltonians which are convex in (x, p). But they are not true in all cases where H is merely Lipschitz continuous in x.

Local solutions to the Hamiltonian equation have a certain optimality property when H is concave-convex. Suppose for instance that x and p are absolutely continuous functions which satisfy (57) over the whole interval $T = [t_0, t_1]$ (almost everywhere). Defining $a_0 = x(t_0)$ and $a_1 = x(t_1)$ and taking l to be the indicator of this endpoint pair as in (14), we see that x and p satisfy the transversality condition for l, as well as (by virtue of the theorem of §10) the Euler-Lagrange condition for L. Hence by the sufficiency theorem in §8, x minimizes $\int_T L(t, x(t), \dot{x}(t)) dt$ over the class of all arcs having the same endpoints a_0 and a_1. Now the same argument can also be applied relative to any subinterval of T. Thus x is *Lagrange optimal* for L over T, in the sense that on every subinterval I it minimizes the Lagrangian integral on I with respect to the class of all arcs that coincide with x at the beginning and end of I. The same can be argued in terms of p via the duality theorem in §9, and one obtains the following.

THEOREM. *In the convex case, if x and p are absolutely continuous functions satisfying the generalized Hamiltonian equation for t in an interval I, then x is Lagrange optimal for L over I, and p is Lagrange optimal for M over I.*

Another special property in the convex case is that if (x, p) and (x', p') are two Hamiltonian trajectories over I, then the quantity $(x(t)-x'(t)) \cdot (p(t)-p'(t))$ is nondecreasing over I [22].

12. Optimal values and perturbations

The close relationship between a problem (P) of convex type and its dual (P*) extends beyond the sharing of sufficient conditions for optimality. There is also a tie between the two optimal values

(63) $\qquad \inf(P) = \inf_{x \in A} \Phi(x), \quad \inf(P^*) = \inf_{p \in A} \Psi(p).$

The study of these values and how they behave under certain "perturbations" of (P) is the route to determining the necessity of the optimality conditions that have been introduced.

A basic inequality can be derived easily from the definition (43) of M and m and the relations (34), (35), that hold for any conjugate pair of convex functions. For arbitrary $x \in A$ and $p \in A$ one has

(64) $$L(t, x(t), \dot{x}(t)) + M(t, p(t), \dot{p}(t)) \geq \dot{p}(t)x(t) + p(t)\dot{x}(t)$$

for almost every $t \in T$, where equality holds if and only if $(\dot{p}(t), p(t)) \in \partial L(t, x(t), \dot{x}(t))$. At the same time

(65) $$l(x(t_0), x(t_1)) + m(p(t_0), p(t_1)) \geq p(t_0)x(t_0) - p(t_1)x(t_1),$$

where equality holds if and only if $(p(t_0), -p(t_1)) \in \partial l(x(t_0), x(t_1))$. Integrating (64) over the interval T and adding (65), we get

(66) $\Phi(x) + \Psi(p) \geq 0$ *for all* $x \in A$, $p \in A$, *with*

equality $\Leftrightarrow$ x *and* p *satisfy the optimality conditions.*

Or do we? There is a slight flaw in the argument, connected with the extended definition of the integrals of L and M as $\pm\infty$. The inequality (66) is quite valid if the convention $\infty - \infty = \infty$ is used on the left side, but the case $\infty - \infty$ could conceivably arise even when (64) and (65) were true with equality, and then there would be strict inequality in (66) despite the optimality conditions being satisfied.

To get around this, a minor assumption must be added. Let L^∞ and L^1 denote the spaces of R^n-valued functions on T which are essentially bounded, or respectively, summable, and define

(67) $$I_L(x, v) = \int_T L(t, x(t), v(t))dt \quad \text{for} \quad (x, v) \in L^\infty \times L^1.$$

ASSUMPTION 5. *The functional I_L is proper on $L^\infty \times L^1$ and bounded below on bounded sets.*

This is satisfied in particular if $\Phi(x) < \infty$ for some $x \in A$ and the Hamiltonian upper boundedness condition holds. Assumption 5 is equivalent in the convex case to the same condition on I_M (hence it is really symmetric in character between (P) and (P*)), and it is also equivalent to:

$$\exists (x, v) \in L^\infty \times L^1 \text{ with } I_L(x, v) < \infty,$$

and

$$\exists (p, r) \in L^\infty \times L^1 \text{ with } I_M(p, r) < \infty.$$

It implies that $\Phi(x)$ and $\Psi(p)$ are never $-\infty$, so the question of $\infty - \infty$ never arises in (66). An important conclusion can then be drawn by rewriting (66) in the form $\Phi(x) \geq -\Psi(p)$.

DUALITY THEOREM 2. *The inequality* $\inf(P) \geq -\inf(P^*)$ *holds for problems of convex type. For* $\min(P) = -\min(P^*)$ *to hold with attainment at* $x \in A$ *and* $p \in A$ *respectively, it is necessary and sufficient that* x *and* p *satisfy the optimality* *conditions*.

The dual of a minimization problem is customarily expressed as a maximization problem, and of course

$$-\inf(P^*) = \sup_{p \in A} \{-\Psi(p)\} .$$

Rather than speaking of the maximization of $-\Psi$ in the present case, we prefer to keep the exact symmetry reflected in the optimality conditions.

The theorem yields an important clue about the circumstances in which the optimality conditions, as stated, are *necessary*.

COROLLARY. *Suppose* $\inf(P) = -\min(P^*)$. *Then* $x \in A$ *furnishes the minimum in* (P) *if and only if it satisfies the optimality conditions in association with some* $p \in A$.

The challenge laid down by this result is to find conditions guaranteeing that $\inf(P) = -\min(P^*)$. An approach can be made through the analysis of the functional

$$\varphi(y, a) = \inf_{x \in A} \left\{ \int_T L(t, x(t)+y(t), \dot{x}(t))dt + l(x(t_0)+a, x(t_1)) \right\} \text{ for } (y, a) \in L^\infty \times R^n .$$

This gives the optimal value in a problem which is like (P) but depends on y and a as parameters (perturbations); clearly $\varphi(0, 0) = \inf(P)$. It is readily seen that φ is convex when (P) is of convex type. Every continuous linear functional on A can be represented in the form $p \mapsto \langle p, (y, a) \rangle$ with

$$(68) \qquad \langle p, (y, a) \rangle = \int_T \dot{p}(t) \cdot y(t) dt + p(0) \cdot a ,$$

so the space $L^\infty \times R^n$ can be identified with A^*. Each $p \in A$ also defines a continuous linear functional $(y, a) \mapsto \langle p, (y, a) \rangle$ on $L^\infty \times R^n$, and it turns out that to have $\inf(P) = -\min(P^*)$ with attainment at p, it is necessary and sufficient that $p \in \partial\varphi(0, 0)$, or in other words,

$$\varphi(y, a) \geq \varphi(0, 0) + \langle p, (y, a) \rangle \text{ for all } (y, a) \in L^\infty \times R^n .$$

This result provides, on the one hand, an interpretation of what the adjoint arc means for (P) itself: it gives coefficients measuring the differential effects of certain perturbations of (P). In particular, if φ happens to be differentiable at $(0, 0)$, one has $p = \nabla\varphi(0, 0)$ in the sense of the pairing (68).

On the other hand, this result reduces the question of whether $\inf(P) = -\min(P^*)$ to the question of the existence of $p \in A$ such that $p \in \partial\varphi(0, 0)$. Such a subgradient p corresponds to a kind of supporting hyperplane to the convex set in $(L^\infty \times R^n) \times R$ which is the epigraph of φ, and so the existence can presumably be obtained from some separation theorem of convex analysis under conditions on L and

l that imply the epigraph has a nonempty interior whose projection on $L^\infty \times R^n$ contains $(0, 0)$.

But there is a catch. With some effort the interiority can be achieved in terms of the topology of $L^\infty \times R^n$ corresponding to the L^∞-norm, but the space of continuous linear functionals in this topology is $(L^\infty)^* \times R^n \simeq A^{**}$, not just A. Thus there is a danger that the supporting hyperplane obtained through separation theory might not be of the form (67), and then it would do no good.

13. Necessity and duality

A crucial restriction must be made to get around the obstacle just explained, and it is dual to the kind of restriction mentioned in §7 in connection with the existence of solutions $x \in A$ for (P).

The *Hamiltonian lower boundedness condition* is satisfied if for each $x \in R^n$ and $\beta \in R$ there is a summable function $\theta : T \mapsto R$ such that

$$H(t, x, p) > \theta(t) \quad \text{for all} \quad t \in T \quad \text{when} \quad |p| \leq \beta.$$

For problems of convex type, this is just the Hamiltonian upper boundedness condition on the dual Lagrangian H' discussed in §10, so it is clearly related to the existence of solutions $p \in A$ for (P*). In particular it requires $H > -\infty$ everywhere.

A concave-convex Hamiltonian satisfies the lower boundedness condition if and only if for every $x \in L^\infty$ there exists $v \in L^1$ with $I_L(x, v)$ finite (where I_L is the functional in (67)). It satisfies both the lower and upper boundedness conditions if and only if $H(t, x, p)$ is a finite, summable function of $t \in T$ for each $(x, p) \in R^n \times R^n$. (See [23, §2] for these and other equivalences.)

The Hamiltonian lower boundedness condition implies for problems of convex type that the epigraph of the functional φ in §12 is of finite codimension and has a nonempty interior relative to its affine hull; furthermore, all subgradients of φ must belong to A, not just A^{**}. This was proved in [27]. The only thing left to be desired is a condition implying that $(0, 0)$ is in the projection on $L^\infty \times R^n$ of the epigraph of φ. This amounts to an attainability condition on the implicit constraints imposed by L and l.

The sets C_L and C_l defined by

$$C_L = \left\{(x_0, x_1) \mid \exists x \in A \text{ with } \int_T L(t, x(t), \dot{x}(t)) dt < \infty, \right.$$

$$\left. x(t_0) = x_0 \text{ and } x(t_1) = x_1 \right\},$$

$$C_l = \{(x_0, x_1) \mid l(x_0, x_1) < \infty\}$$

obviously have the property that

$$C_L \cap C_l \neq \emptyset \Leftrightarrow \exists x \in A \text{ with } \Phi(x) < \infty .$$

The *attainability condition* for (P) is the slightly stronger property that ri $C_L \cap$ ri $C_l \neq \emptyset$, where "ri" denotes the relative interior of a convex set (its interior with respect to its affine hull, see [20, §6]). It is certainly satisfied if C_l is all of $R^n \times R^n$ and $C_L \neq \emptyset$, or if C_l consists of a single point lying in the relative interior of C_L. (In [23] the definition of C_L is a bit different but shown to be equivalent to the one here.) The attainability condition for (P*) is the same thing in terms of M and m.

DUALITY THEOREM 3. *For problems of convex type, the following hold.*

(a) If the attainability condition for (P) *is satisfied and* H *has the lower boundedness property, then* inf(P) = -min(P*) < ∞ *, and for* $x \in A$ *to furnish the minimum in* (P) *it is necessary (as well as sufficient) that* x *satisfy the optimality conditions (in association with some* $p \in A$ *).*

(b) If the attainability condition for (P*) *is satisfied and* H *has the upper boundedness property, then* min(P) = -inf(P*) > -∞ *, and for* $p \in A$ *to furnish the minimum in* (P*) *it is necessary (as well as sufficient) that* p *satisfy the optimality conditions (in association with some* $x \in A$ *).*

This is the main theorem of [23]. Note that *(b)* is an existence theorem for (P), just as *(a)* is an existence theorem for (P*).

The attainability condition for (P*) can be translated into a growth condition on the convex functional Φ in (P) (see [23]). A condition on L implying in the autonomous case that the sets C_L and C_M in the attainability conditions are non-empty and project onto all of R^n in either argument, regardless of the choice of the interval T, may be found in [28, p. 151].

The most interesting feature is the duality between the existence of solutions to one problem and the necessity of the optimality conditions in the other. The two are closely connected, for better or for worse. The "worse" aspect is that, while the Hamiltonian lower boundedness condition is welcome enough as a burden en route to the existence of solutions to (P*), it has the unwanted effect of eliminating the possibility of real state constraints in (P). Indeed, such constraints appear in the implicit form $x(t) \in X(t)$ almost everywhere, where

$$X(t) = \{x \in R^n \mid \exists v \in R^n \text{ with } L(t, x, v) < \infty\} ,$$

and the lower boundedness condition implies via (29) that $X(t) = R^n$ for all $t \in T$.

However, the fact that state constraints become involved in this way is quite natural, when one thinks about it. The optimality conditions that have been derived for standard kinds of control problems with state constraints typically include multipliers (dual variables) that can jump at times t when $x(t)$ touches the boundary of the state constraint region. This suggests that an adequate treatment of such problems would involve adjoint arcs p that might not be continuous. Since p is required to be absolutely continuous in problem (P*) and the optimality conditions, it is no wonder that in order to get the necessity of the optimality conditions we have had to impose a restriction that eliminates state constraints.

Where does this leave us in our desire to have a theory applicable also to problems with state constraints? A fundamental extension of the framework is needed. The optimality conditions must be generalized to admit arcs p in a larger space than A, and the natural choice turns out to be the space of arcs of bounded variation. If duality is still to play a role, the formulation of (P*) must also be extended to this space. Thus we must decide what $\Psi(p)$ should mean for an arc of bounded variation. But symmetry demands that whatever is done for p should be done for x. Both (P) and (P*) should therefore be in terms of arcs of bounded variation. The hope is that the extended problems will be just "closures" of the original problems in some sense, and it can be left to the optimality conditions themselves to tell us whether a particular solution arc or adjoint arc must actually be absolutely continuous.

14. Arcs of bounded variation

The treatment of state constraints has led us to the question of how to generalize the Hamiltonian equations and the functional Φ from A to the space B of R^n-valued functions of bounded variation on the interval T. An answer that takes care of both of these needs is found in making the right generalization of ordinary differential (contingent) equations of the form

(69) $\qquad \dot{z}(t) \in C(t, z(t))$ almost everywhere, $z \in A$,

to the case of $z \in B$. The Hamiltonian equation is of such type, and the study of the functional $\int_T L(t, x(t), \dot{x}(t)) dt$ can be reduced if necessary to the study of (69) for $z(t) = (x(t), x_0(t))$ and

(70) $\qquad C(t, z(t)) =$ epigraph of $L(t, x(t), \cdot)$.

Certain simplifications are possible for problems of convex type, but even in the general case it is reasonable to assume at the very least that $C(t, z)$ is a closed convex set for each $t \in T$ and $z \in R^n$ (possibly empty for z belonging to some "forbidden region"), and furthermore that the graph $\Gamma(t) = \{(z, w) \mid w \in C(t, z)\}$ is closed and depends measurably on t, that is, Γ is a measurable multifunction. (For (70), the convexity of $C(t, z)$ corresponds to Assumption 4 in §7.)

For a nonempty closed convex set C, the *recession cone* of C, denoted by 0^+C, is the "limit" of $\lambda C = \{\lambda w \mid w \in C\}$ as $\lambda \to 0^+$ (see [20, §8]). It reduces to $\{0\}$ if and only if C is bounded. The basic idea for extending (69) is the following. Each $z \in B$ corresponds to an R^n-valued Borel measure dz on T, and there always exists a nonnegative Borel measure on T with respect to which both dz and the Lebesgue measure dt are absolutely continuous. The latter can be expressed as $d\tau$ for a real valued function τ on T which is increasing (hence also of bounded variation). If $d\tau$ is absolutely continuous with respect to dt, we can use Radon-Nikodym derivatives to write (69) equivalently as

$$(71) \qquad \frac{dz}{d\tau}(t) \in \frac{dt}{d\tau}(t) \cdot C(t, z(t)) \quad \text{almost everywhere} \quad (d\tau),$$

where $(dt/d\tau)(t) > 0$ almost everywhere $(d\tau)$. If dt is not absolutely continuous with respect to dt, this is reflected by having merely $(dt/d\tau)(t) \geq 0$ almost everywhere $(d\tau)$. The generalization consists essentially of adopting (71) as the replacement for (69) in this case with the right side interpreted as $0^+C(t, z(t))$ when $(dt/d\tau)(t) = 0$.

What one gets is actually independent of the particular choice of $d\tau$. It is equivalent to augmenting the earlier equation (69) (which still makes sense - the derivative $\dot{z} = dz/dt$ does exist, but unless z is absolutely continuous it will not be the integral of $\dot{z}dt$) by a special condition on the singular part of dz:

$$(72) \qquad \frac{dz}{d\tau}(t) - \frac{dz}{dt}(t)\frac{dt}{d\tau}(t) \in 0^+C(t, z(t)) \quad \text{almost everywhere} \quad (d\tau).$$

For the generalized "equation" (69) plus (72), the notation

$$dz(t) \in C(t, z(t))dt$$

seems appropriate.

But there are some wrinkles to be ironed out. In (72) the left side is measurable with respect to $d\tau$, not just dt, so something other than Lebesgue measurability should apparently be demanded of the multifunction $t \mapsto 0^+C(t, z(t))$ as well. The possible jumps in z also cause a problem. Besides $z(t)$, one has the limits $z(t+)$ and $z(t-)$, and there can be a countable infinity of points t at which these might not all agree. At such a point, (72) gives the jump condition

$$z(t+) - z(t-) \in 0^+C(t, z(t)),$$

but there is some doubt about whether $z(t)$ is really the correct thing to have on the right side or $z(t+)$ or $z(t-)$ (or both), particularly since we may just want to forget about $z(t)$ itself and identify functions of bounded variation which have the same one-sided limits at each point. Another question concerns what $0^+C(t, z)$ should be when $C(t, z) = \emptyset$ but $C(t, z_k) \neq \emptyset$ and $0^+C(t, z_k) \neq \{0\}$ for a sequence of points z_k converging to z.

More work is needed in the general case, but these riddles can be answered in a satisfying manner in the context of the application to the theory of state constraints in problems of convex type, *cf.* [24], [30]. The conditions on the Hamiltonian that replace upper and lower boundedness concern the state constraint set $X(t)$ and the corresponding set

$$P(t) = \{p \in R^n \mid \exists r \in R^n \text{ with } M(t, p, r)\}$$

for the dual problem. These are always convex and have the property that

$$H(t, x, p) = \begin{cases} \text{finite value if } x \in X(t), \ p \in P(t), \\ +\infty \quad \text{if } x \in X(t), \ p \notin \text{cl } P(t), \\ -\infty \quad \text{if and only if } x \notin X(t). \end{cases}$$

The case treated in [30] is the one where $X(t)$ and $P(t)$ have nonempty interiors which depend "continuously" on t, and $H(t, x, p)$ is summable in t over finite intervals during which x and p are in the interiors of $X(t)$ and $P(t)$. In the framework of the development outlined for the proof of theorem in the preceding section, the functional $\varphi(y, a)$ is restricted to $C \times R^n$ instead of $L^\infty \times R^n$, so the dual space can be identified with B.

The extended Hamiltonian equation is in terms of

$$C(t, x, p) = \{(v, r) \mid (-r, v) \in \partial H(t, x, p)\},$$

and if $X(t)$ and $P(t)$ are closed the singular part (72) reduces to a condition in terms of

$$0^+ C(t, x, p) = N_{P(t)}(p) \times N_{X(t)}(x),$$

where $N_{X(t)}$ and $N_{P(t)}$ are the normal cones defined in (38). Results on duality, existence, and necessary and sufficient conditions are obtained, much like those above. Furthermore, solutions to the extended problems in B can be characterized as limits of minimizing sequences for the original problems in A. See [24], [30], for details.

15. Problems over an infinite horizon

There is considerable interest among mathematical economists in problems of convex type with the interval T unbounded, for example, $T = [0, \infty)$. Typically the Lagrangian is of the form

$$L(t, x, v) = -e^{\rho t} U(x, v),$$

where U is a concave "utility" function and ρ is the "discount rate".

When $\rho = 0$, the Hamiltonian is independent of t and expressed by

$$H(x, p) = \sup_{v \in R^n} \{p \cdot v + U(x, v)\} .$$

Since H is concave in x and convex in p, it may well have a saddle point $(\bar{x}, \bar{p})$ in the minimax sense:

$$H(x, \bar{p}) \leq H(\bar{x}, \bar{p}) \leq H(\bar{x}, p) \quad \text{for all} \quad x, p .$$

It has been demonstrated in [25] that if H happens to be *strictly* concave in x and *strictly* concave in p in a neighborhood of $(\bar{x}, \bar{p})$, then $(\bar{x}, \bar{p})$ is also a saddle point for the Hamiltonian equation in the sense that the term "saddle point" is used for dynamical systems. More specifically, in a neighborhood of $(\bar{x}, \bar{p})$ the Hamiltonian trajectories $(x(t), p(t))$ that tend to $(\bar{x}, \bar{p})$ as $t \to +\infty$ make up an n-dimensional manifold K_+ in R^{2n}, while those that tend to $(\bar{x}, \bar{p})$ as $t \to -\infty$ form a similar manifold K_- with $K_+ \cap K_- = \{(\bar{x}, \bar{p})\}$. The trajectories in K_+ have a certain natural optimality property over intervals $[t_0, \infty)$, while those in K_- have such a property for $(-\infty, t_1]$.

These results have been obtained through application of the duality theory described here (without getting involved with state constraints). A kind of extension to the case where $\rho > 0$ is carried out in [31].

OTHER EXTENSIONS OF THE THEORY. The duality between (P) and (P*) has been generalized by Barbu [1], [2], [3], [4], [5], to problems where the states $x(t)$ are not in R^n but an infinite-dimensional Hilbert space. Some applications to systems governed by partial differential equations are thereby covered. For another case corresponding to partial differential equations, namely where the interval T is replaced by a region Ω in R^k and $\dot{x}$ by Dx for some operator D, see the book of Ekeland and Temam [17]. Bismut [6] has applied the duality theory to problems in stochastic optimal control.

References

[1] Viorel Barbu, "Convex control problems of Bolza in Hilbert spaces", *SIAM J. Control* 13 (1975), 754-771.

[2] Viorel Barbu, "On the control problem of Bolza in Hilbert spaces", *SIAM J. Control* 13 (1975), 1062-1076.

[3] V. Barbu, "Convex control problems for linear differential systems of retarded type", *Ricerche Mat.* 26 (1977), 3-26.

[4] Viorel Barbu, "Constrained control problems with convex cost in Hilbert space", *J. Math. Anal. Appl.* 56 (1976), 502-528.

[5] V. Barbu, "On convex control problems on infinite intervals", submitted.

[6] Jean-Michel Bismut, "Conjugate convex functions in optimal stochastic control", *J. Math. Anal. Appl.* **44** (1973), 384-404.

[7] Charles Castaing, "Sur les équations différentielles multivoques", *C.R. Acad. Sci. Paris Sér. A-B* **263** (1966), A63-A66.

[8] Lamberto Cesari, "Existence theorems for weak and usual optimal solutions in Lagrange problems with unilateral constraints. I", *Trans. Amer. Math. Soc.* **124** (1966), 369-412.

[9] Frank H. Clarke, "Admissible relaxation in variational and control problems", *J. Math. Anal. Appl.* **51** (1975), 557-576.

[10] Frank H. Clarke, "Generalized gradients and applications", *Trans. Amer. Math. Soc.* **205** (1975), 247-262.

[11] Frank H. Clarke, "The Euler-Lagrange differential inclusion", *J. Differential Equations* **19** (1975), 80-90.

[12] Frank H. Clarke, "La condition hamiltonienne d'optimalité", *C.R. Acad. Sci. Paris Sér. A-B* **280** (1975), A1205-A1207.

[13] Frank H. Clarke, "The generalized problem of Bolza", *SIAM J. Control Optimization* **14** (1976), 682-699.

[14] Frank H. Clarke, "Necessary conditions for a general control problem", *Calculus of Variations and Control Theory* (Symposium, University Wisconsin, Madison, Wisconsin, 1975, 257-278. Academic Press, New York, San Francisco, London, 1976).

[15] Frank H. Clarke, "The maximum principle under minimal hypothesis", *SIAM J. Control Optimization* **14** (1976), 1078-1091.

[16] F.H. Clarke, "Generalized gradients of Lipschitz functionals", submitted.

[17] Ivar Ekeland and Roger Temam, *Convex Analysis and Variational Problems* (Studies in Mathematics and its Applications, 1. North-Holland, Amsterdam, Oxford, New York, 1976).

[18] W. Fenchel, "On conjugate convex functions", *Canad. J. Math.* **1** (1949), 73-77.

[19] Czeslaw Olech, "Existence theorems for optimal problems with vector-valued cost function", *Trans. Amer. Math. Soc.* **136** (1969), 159-180.

[20] R. Tyrrell Rockafellar, *Convex Analysis* (Princeton Mathematical Series, **28**. Princeton University Press, Princeton, New Jersey, 1970).

[21] R.T. Rockafellar, "Conjugate convex functions in optimal control and the calculus of variations", *J. Math. Anal. Appl.* **32** (1970), 174-222.

[22] R. Tyrrell Rockafellar, "Generalized Hamiltonian equations for convex problems of Lagrange", *Pacific J. Math.* **33** (1970), 411-427.

[23] R.T. Rockafellar, "Existence and duality theorems for convex problems of Bolza", *Trans. Amer. Math. Soc.* **159** (1971), 1-40.

[24] R. Tyrrell Rockafellar, "State constraints in convex control problems of Bolza", *SIAM J. Control* **10** (1972), 691-715.

[25] R.T. Rockafellar, "Saddle points of Hamiltonian systems in convex problems of Lagrange", *J. Optimization Theory Appl.* **12** (1973), 367-390.

[26] R. Tyrrell Rockafellar, *Conjugate Duality and Optimization* (Conference Board of the Mathematical Sciences, Regional Conference Series in Applied Math., **16**. Society for Industrial and Applied Mathematics, Philadelphia, 1974).

[27] R. Tyrrell Rockafellar, "Existence theorems for general control problems of Bolza and Lagrange", *Advances in Math.* **15** (1975), 312-333.

[28] R. Tyrrell Rockafellar, "Semigroups of convex bifunctions generated by Lagrange problems in the calculus of variations", *Math. Scand.* **36** (1975), 137-158.

[29] R. Tyrrell Rockafellar, "Integral functionals, normal integrands and measurable selections", *Nonlinear Operators and the Calculus of Variations* (Lecture Notes in Mathematics, **543**, 157-207. Springer-Verlag, Berlin, Heidelberg, New York. 1976)

[30] R. Tyrrell Rockafellar, "Dual problems of Lagrange for arcs of bounded variation", *Calculus of Variations and Control Theory* (Symposium, University Wisconsin, Madison, 1975, 155-192. Academic Press, New York, San Francisco, London, 1976).

[31] R. Tyrrell Rockafellar, "Saddle points of Hamiltonian systems in convex Lagrange problems having a nonzero discount rate. Hamiltonian Dynamics in economics", *J. Econom. Theory* **12** (1976), 71-113.

[32] D.H. Wagner, "Survey of measurable selection theorems", *SIAM J. Control Optimization* **15** (1977), 859-903.

[33] J. Warga, "Relaxed variational problems", *J. Math. Anal. Appl.* **4** (1962), 111-128.

QA
3
L28
v.680

DEC 11 1978